CAMBRIDGE LIBRARY COLLECTION

Books of enduring scholarly value

Physical Sciences

From ancient times, humans have tried to understand the workings of the world around them. The roots of modern physical science go back to the very earliest mechanical devices such as levers and rollers, the mixing of paints and dyes, and the importance of the heavenly bodies in early religious observance and navigation. The physical sciences as we know them today began to emerge as independent academic subjects during the early modern period, in the work of Newton and other 'natural philosophers', and numerous sub-disciplines developed during the centuries that followed. This part of the Cambridge Library Collection is devoted to landmark publications in this area which will be of interest to historians of science concerned with individual scientists, particular discoveries, and advances in scientific method, or with the establishment and development of scientific institutions around the world.

Personal Recollections, from Early Life to Old Age

These *Personal Recollections* contain the memoirs and a selection of the correspondence of the nineteenth-century polymath Mary Somerville (1780–1872). The book was first published in 1873, a year after Mary's death, by her daughter Martha, who wrote brief introductions to the text. Mary Somerville is best known for her pioneering scientific publications which include her translation of Laplace's *Mécanique Céleste* (1831: also resissued in this series); *On the Connection of the Physical Sciences* (1834); *Physical Geography* (1848); and *On Molecular and Microscopic Science* (1869). Through these publications, Somerville made a lasting contribution to the dissemination of scientific knowledge. Somerville's correspondence deals primarily with her public life, while the memoirs offer insight into her private sphere: the discouragement she faced in pursuit of learning; her passion for women's education and suffrage; family life; and personal faith. Her story is compelling, and her experiences may resonate with many women today.

Cambridge University Press has long been a pioneer in the reissuing of out-of-print titles from its own backlist, producing digital reprints of books that are still sought after by scholars and students but could not be reprinted economically using traditional technology. The Cambridge Library Collection extends this activity to a wider range of books which are still of importance to researchers and professionals, either for the source material they contain, or as landmarks in the history of their academic discipline.

Drawing from the world-renowned collections in the Cambridge University Library, and guided by the advice of experts in each subject area, Cambridge University Press is using state-of-the-art scanning machines in its own Printing House to capture the content of each book selected for inclusion. The files are processed to give a consistently clear, crisp image, and the books finished to the high quality standard for which the Press is recognised around the world. The latest print-on-demand technology ensures that the books will remain available indefinitely, and that orders for single or multiple copies can quickly be supplied.

The Cambridge Library Collection will bring back to life books of enduring scholarly value (including out-of-copyright works originally issued by other publishers) across a wide range of disciplines in the humanities and social sciences and in science and technology.

Personal Recollections, from Early Life to Old Age

With Selections from her Correspondence

Mary Somerville

Edited by Martha Somerville

Cambridge University Press

CAMBRIDGE UNIVERSITY PRESS

Cambridge, New York, Melbourne, Madrid, Cape Town, Singapore,
São Paolo, Delhi, Dubai, Tokyo

Published in the United States of America by Cambridge University Press, New York

www.cambridge.org
Information on this title: www.cambridge.org/9781108013659

This edition first published 1873
This digitally printed version 2010

ISBN 978-1-108-01365-9 Paperback

MARY SOMERVILLE

PERSONAL RECOLLECTIONS,

FROM EARLY LIFE TO OLD AGE,

OF

MARY SOMERVILLE.

WITH

Selections from her Correspondence.

BY HER DAUGHTER, MARTHA SOMERVILLE.

LONDON:
JOHN MURRAY, ALBEMARLE STREET.
1873.

WORKS BY MRS. SOMERVILLE.

THE MECHANISM OF THE HEAVENS. 8vo. 1831.

THE CONNECTION OF THE PHYSICAL SCIENCES. *9th Edition.* Post 8vo. 9*s.* 1858.

PHYSICAL GEOGRAPHY. *6th Edition.* Post 8vo. 9*s.* 1870.

MOLECULAR AND MICROSCOPIC SCIENCE. 2 vols. Post 8vo. 21*s.* 1869.

CONTENTS.

CHAPTER VII.

CHAPTER VIII.

CHAPTER IX.

CHAPTER X.

CHAPTER XI.

CHAPTER XII.

CHAPTER XIII.

CHAPTER XIV.

CHAPTER XV.

CHAPTER XVI.

PERSONAL RECOLLECTIONS

OF

MARY SOMERVILLE.

CHAPTER I.

INTRODUCTION—PARENTAGE—LIFE IN SCOTLAND IN THE LAST CENTURY—EARLY EDUCATION—SCHOOL.

THE life of a woman entirely devoted to her family duties and to scientific pursuits affords little scope for a biography. There are in it neither stirring events nor brilliant deeds to record; and as my Mother was strongly averse to gossip, and to revelations of private life or of intimate correspondence, nothing of the kind will be found in the following pages. It has been only after very great hesitation, and on the recommendation of valued friends, who think that some account of so remarkable and beautiful a character cannot fail to interest the public, that I have resolved to publish some detached Recollections of past times, noted down by my mother during the last years of her life, together with a few letters from eminent men and women, referring almost exclusively to her scientific works. A still smaller number of her own letters have been added, either as illustrating her opinions on events she witnessed, or else as affording some slight idea of her simple and loving disposition.

Few thoughtful minds will read without emotion my mother's own account of the wonderful energy and indomitable perseverance by which, in her ardent thirst for knowledge, she overcame obstacles apparently insurmountable, at a time when women were well-nigh totally debarred from education; and the almost intuitive way in which she entered upon studies of which she had scarcely heard the names, living, as she did, among persons to whom they were utterly unknown, and who disapproved of her devotion to pursuits so different from those of ordinary young girls at the end of the last century, especially in Scotland, which was far more old-fashioned and primitive than England.

Nor is her simple account of her early days without interest, when, as a lonely child, she wandered by the seashore, and on the links of Burntisland, collecting shells and flowers; or spent the clear, cold nights at her window, watching the starlit heavens, whose mysteries she was destined one day to penetrate in all their profound and sublime laws, making clear to others that knowledge which she herself had acquired, at the cost of so hard a struggle.

It was not only in her childhood and youth that my mother's studies encountered disapproval. Not till she became a widow, had she perfect freedom to pursue them. The first person—indeed the only one in her early days—who encouraged her passion for learning was her uncle by marriage, afterwards her father-in-law, the Rev. Dr. Somerville, minister of Jedburgh, a man very much in advance of his century in liberality of thought on all subjects. He was one of the first to discern her rare qualities, and valued her as she deserved; while through life she retained the most grateful affection for him, and confided to him many doubts and difficulties on subjects of the highest importance. Nothing can be more erroneous than the statement, repeated in several obituary notices of my mother, that Mr. Greig (her first husband) aided her in her mathematical and other pursuits. Nearly the contrary was the case. Mr. Greig took no interest in science or literature, and possessed in full the prejudice against learned women which was common at that time. Only on her marriage with my father, my mother at last met with one who entirely sympathised with her, and warmly entered into all her ideas, encouraging her zeal for study to the utmost, and affording her every facility for it in his power. His love and admiration for her were unbounded; he frankly and willingly acknowledged her superiority to himself, and many of our friends can bear witness to the honest pride and gratification which he always testified in the fame and honours she attained.

No one can escape sorrow, and my mother, in the course of her long life, had her full share, but she bore it with that deep feeling of trust in the great goodness of God which formed so marked a feature in her character. She had a buoyant and hopeful spirit, and though her affections were very strong, and she felt keenly, it was ever her nature to turn from the shadows to all that is bright and beautiful in mortal life. She had much to make life pleasant in the great honours universally bestowed upon her; but she found far more in the devoted affection of friends, to say nothing of those whose happy lot it has been to live in close and loving intercourse with so noble and gentle a spirit.

She met with unbounded kindness from men of science of all countries, and most profound was her gratitude to them. Modest and unpretending to excess, nothing could be more generous than the unfeigned delight she shewed in recognising the genius and discoveries of others; ever jealous of their fame, and never of her own.

It is not uncommon to see persons who hold in youth opinions in advance of the age in which they live, but who at a certain period seem to crystallise, and lose the faculty of comprehending and accepting new ideas and theories; thus remaining at last as far behind, as they were once in advance of public opinion. Not so my mother, who was ever ready to hail joyfully any new idea or theory, and to give it honest attention, even if it were at variance with her former convictions. This quality she never lost, and it enabled her to sympathise with the younger generation of philosophers, as she had done with their predecessors, her own contemporaries.

Although her favourite pursuit, and the one for which she had decidedly most aptitude, was mathematics; yet there were few subjects in which she did not take interest, whether in science or literature, philosophy or politics. She was passionately fond of poetry, her especial favourites being Shakespeare and Dante, and also the great Greek dramatists, whose tragedies she read fluently in the original, being a good classical scholar. She was very fond of music, and devoted much time to it in her youth, and she painted from nature with considerable taste. The latter was, perhaps, the recreation in which she most delighted, from the opportunity it afforded her of contemplating the wonderful beauty of the world, which was a never-failing source of intense enjoyment to her, whether she watched the changing effects of light and shade on her favourite Roman Campagna, or gazed, enchanted, on the gorgeous sunsets on the bay of Naples, as she witnessed them from her much-loved Sorrento, where she passed the last summers of her life. All things fair were a joy to her—the flowers we brought her from our rambles, the sea-weeds, the wild birds she saw, all interested and pleased her. Everything in nature spoke to her of that great God who created all things, the grand and sublimely beautiful as well as the exquisite loveliness of minute objects. Above all, in the laws which science unveils step by step, she found ever renewed motives for the love and adoration of their Author and Sustainer. This fervour of religious feeling accompanied her through life, and very early she shook off all that was dark and narrow in the creed of her first instructors for a purer and a happier faith.

It would be almost incredible were I to describe how much my mother contrived to do in the course of the day. When my sister and I were small children, although busily engaged in writing for the press, she used to teach us for three hours every morning, besides managing her house carefully, reading the newspapers (for she always was a keen, and, I must add, a liberal politician), and the most important new books on all subjects, grave and gay. In addition to all this, she freely visited and received her friends. She was, indeed, very fond of society, and did not look for transcendent talent in those with whom she associated, although no one appreciated it more when she found it. Gay and cheerful company was a pleasant relaxation after a hard day's work. My mother never introduced scientific or learned subjects into general conversation. When they were brought forward by others, she talked simply and naturally about them, without the slightest pretension to superior knowledge. Finally, to complete the list of her accomplishments, I must add that she was a remarkably neat and skilful needlewoman. We still possess some elaborate specimens of her embroidery and lace-work.

Devoted and loving in all the relations of life, my mother was ever forgetful of self. Indulgent and sympathising, she never judged others with harshness or severity; yet she could be very angry when her indignation was aroused by hearing of injustice or oppression, of cruelty to man or beast, or of any attack on those she loved. Rather timid and retiring in general society, she was otherwise fearless in her quiet way. I well remember her cool composure on some occasions when we were in great danger. This she inherited from her father, Admiral Sir William Fairfax, a gallant gentleman who distinguished himself greatly at the battle of Camperdown.*

My mother speaks of him as follows among her "Recollections," of which I now proceed to place some portions before the reader.

---

* Sir William Fairfax was the son of Joseph Fairfax, Esq., of Bagshot, in the county of Surrey, who died in 1783, aged 77, having served in the army previous to 1745. It is understood that his family was descended from the Fairfaxes of Walton, in Yorkshire, the main branch of which were created Viscounts Fairfax of Emly, in the peerage of Ireland (now extinct), and a younger branch Barons Fairfax of Cameron, in the peerage of Scotland. Of the last-named was the great Lord Fairfax, Commander-in-Chief of the armies of the Parliament, 1645—50, whose title is now held by the eleventh Lord Fairfax, a resident in the United States of America.

My father was very good looking, of a brave and noble nature, and a perfect gentleman both in appearance and character. He was sent to sea as midshipman at ten years of age, so he had very little education; but he read a great deal, chiefly history and voyages. He was very cool, and of instant resource in moments of danger.

One night, when his little vessel had taken refuge with many others from an intensely violent gale and drifting snow in Yarmouth Roads, they saw lights disappear, as vessel after vessel foundered. My father, after having done all that was possible for the safety of the ship, went to bed. His cabin door did not shut closely, from the rolling of the ship, and the man who was sentry that night told my mother years afterwards, that when he saw my father on his knees praying, he thought it would soon be all over with them; then seeing him go to bed and fall asleep, he felt no more fear. In the morning the coast was strewed with wrecks. There were no life-boats in those days; now the lives of hundreds are annually saved by the noble self-devotion of British sailors.

My mother was the daughter of Samuel Charters, Solicitor of the Customs for Scotland, and his wife

Christian Murray, of Kynynmont, whose eldest sister married the great grandfather of the present Earl of Minto. My grandmother was exceedingly proud and stately. She made her children stand in her presence. My mother, on the contrary, was indulgent and kind, so that her children were perfectly at ease with her. She seldom read anything but the Bible, sermons, and the newspaper. She was very sincere and devout in her religion, and was remarkable for good sense and great strength of expression in writing and conversation. Though by no means pretty, she was exceedingly distinguished and ladylike both in appearance and manners.

My father was constantly employed, and twice distinguished himself by attacking vessels of superior force. He captured the first, but was overpowered by the second, and being taken to France, remained two years a prisoner on parole, when he met with much kindness from the Choiseul family. At last he was exchanged, and afterwards was appointed lieutenant on board a frigate destined for foreign service. I think it was the North American station, for the war of Independence was not over till the beginning of 1783. As my mother knew that my father would be absent for some years, she accompanied him to London, though so near her confinement that in returning home she had just time to arrive at the manse of Jedburgh, her sister Martha Somerville's* house, when I was born, on the 26th December, 1780. My mother was dangerously ill, and my aunt, who was about to wean her second daughter Janet, who married General Henry Elliot, nursed me till a wetnurse could be found. So I was born in the house of my future husband, and nursed by his mother—a rather singular coincidence.

During my father's absence, my mother lived with great economy in a house not far from Burntisland which belonged to my grandfather, solely occupied with the care of her family, which consisted of her eldest son Samuel, four or five years old, and myself. One evening while my brother was lying at play on the floor, he called out, " O, mamma, there's the moon rinnin' awa." It was the celebrated meteor of 1783.

Some time afterwards, for what reason I do not know, my father and mother went to live for a short time at Inveresk, and thence returned to Burntisland, our permanent home.

* * * * *

[This place, in which my mother's early life was spent, exercised so much influence on her life and pursuits,

* Wife of the Rev. Dr. Thomas Somerville, minister of Jedburgh, already mentioned (p. 2). Dr. Somerville was author of Histories of Queen Anne and of William and Mary, and also of an autobiography.

that I am happy to be able to give the description of it in her own words.

Burntisland was then a small quiet seaport town with little or no commerce, situated on the coast of Fife, immediately opposite to Edinburgh. It is sheltered at some distance on the north by a high and steep hill called the Bin. The harbour lies on the west, and the town ended on the east in a plain of short grass called the Links, on which the towns-people had the right of pasturing their cows and geese. The Links were bounded on each side by low hills covered with gorse and heather, and on the east by a beautiful bay with a sandy beach, which, beginning at a low rocky point, formed a bow and then stretched for several miles to the town of Kinghorn, the distant part skirting a range of high precipitous crags.

Our house, which lay to the south of the town, was very long, with a southern exposure, and its length was increased by a wall covered with fruit-trees, which concealed a courtyard, cow-house, and other offices. From this the garden extended southwards, and ended in a plot of short grass covering a ledge of low black rocks washed by the sea. It was divided into three parts by narrow, almost unfrequented, lanes. These gardens yielded abundance of common fruit and vegetables, but the warmest and best exposures were always devoted to flowers. The garden next to the house was bounded on the south by an ivy-covered wall hid by a row of old elm trees, from whence a steep mossy bank descended to a flat plot of grass with a gravel walk and flower borders on each side, and a broad gravel walk ran along the front of the house. My mother was fond of flowers, and prided herself on her moss-roses, which flourished luxuriantly on the front of the house; but my father, though a sailor, was an excellent florist. He procured the finest bulbs and flower seeds from Holland, and kept each kind in a separate bed.

The manners and customs of the people who inhabited this pretty spot at that time were exceedingly primitive.

Upon the death of any of the townspeople, a man went about ringing a bell at the doors of the friends and acquaintances of the person just dead, and, after calling out "Oyez!" three times, he announced the death which had occurred. This was still called by the name of the Passing-bell, which in Catholic times invited the prayers of the living for the spirit just passed away.

There was much sympathy and kindness shown on these occasions; friends always paid a visit of condolence to the afflicted, dressed in black. The gude wives in Burntisland thought it respectable to provide dead-clothes for themselves and the "gude man," that they might have a decent funeral. I once saw a set of grave-clothes nicely folded up, which consisted of a long shirt and cap of white flannel, and a shroud of fine linen made of yarn, spun by the gude wife herself. I did not like that gude wife; she was purse-proud, and took every opportunity of treating with scorn a poor neighbour who had had a *misfortune*, that is, a child by her husband before marriage, but who made a very good wife. Her husband worked in our garden, and took our cow to the Links to graze. The wife kept a little shop, where we bought things, and she told us her neighbour had given her "mony a sair greet"—that is, a bitter fit of weeping.

The howdie, or midwife, was a person of much consequence. She had often to go far into the country, by day and by night, riding a cart-horse. The neighbours used to go and congratulate the mother, and, of course, to admire the baby. Cake and caudle were handed round, caudle being oat-meal gruel, with sugar, nutmeg, and white wine. In the poorest class, hot ale and "scons" were offered.

Penny-weddings were by no means uncommon in my young days. When a very poor couple were going to be married, the best man, and even the bridegroom himself, went from house to house, asking for small sums to enable them to have a wedding supper, and pay the town fiddler for a dance; any one was admitted who paid a penny. I recollect the prisoners in the Tolbooth letting down bags from the prison windows, begging for charity. I do not remember any execution taking place.

Men and old women of the lower classes smoked tobacco in short pipes, and many took snuff—even young ladies must have done so; for I have a very pretty and quaint gold snuff-box which was given to my grandmother as a marriage present. Licensed beggars, called "gaberlunzie men," were still common. They wore a blue coat, with a tin badge, and wandered about the country, knew all that was going on, and were always welcome at the farm-houses, where the gude wife liked to have a crack (gossip) with the blue coat, and, in return for his news, gave him dinner or supper, as might be. Edie Ochiltree is a perfect specimen of this extinct race. There was another species of beggar, of yet higher antiquity. If a man were a cripple, and poor, his relations put him in a hand-barrow, and wheeled him to their next neighbour's door, and left him there. Some one came out, gave him oat-cake or peasemeal bannock, and then wheeled him to the next door; and in this way, going from house to house, he obtained a fair livelihood.

My brother Sam lived with our grandfather in Edinburgh, and attended the High School, which was in the old town, and, like other boys, he was given pennies to buy bread; but the boys preferred oysters, which they bought from the fishwives, the bargain being, a dozen oysters for a halfpenny, and a kiss for the thirteenth. These fishwives and their husbands were industrious, hard-working people, forming a community of their own in the village of Newhaven, close to the sea, and about two miles from Edinburgh. The men were exposed to cold, and often to danger, in their small boats, not always well-built nor fitted for our stormy Firth. The women helped to land and prepare the fish when the boats came in, carried it to town for sale in the early morning, kept the purse, managed the house, brought up the children, and provided food and clothing for all. Many were rich, lived well, and sometimes had dances. Many of the young women were pretty, and all wore—and, I am told, still wear —a bright-coloured, picturesque costume. Some young men, amongst others a cousin of my own, who attempted to intrude into one of these balls, got pelted with fish offal by the women. The village smelt strongly of fish, certainly; yet the people were very clean personally. I recollect their keeping tame gulls, which they fed with fish offal.

Although there was no individual enmity between the boys of the old and of the new or aristocratic part of Edinburgh, there were frequent battles, called "bickers," between them, in which they pelted each other with stones. Sometimes they were joined by bigger lads, and then the fight became so serious that the magistrates sent the city guard—a set of old men with halberds and a quaint uniform—to separate them; but no sooner did the guard appear, than both parties joined against them.

Strings of wild geese were common in autumn, and I was amused on one occasion to see the clumsy tame fat geese which were feeding on the Links rise in a body and try to follow the wild ones.

As the grass on the plot before our house did not form a fine even turf, the ground was trenched and sown with good seed, but along with the grass a vast crop of thistles and groundsel appeared, which attracted quantities of goldfinches, and in the early mornings I have seen as many as sixty to eighty of these beautiful birds feeding on it.

My love of birds has continued through life, for only two years ago, in my extreme old age, I lost a pet mountain sparrow, which for eight years was my constant companion: sitting on my shoulder, pecking at my papers, and eating out of my mouth; and I am not ashamed to say I felt its accidental death very much.

Before the grass came up on this plot of ground, its surface in the evening swarmed with earthworms, which instantly shrank into their holes on the approach of a foot. My aunt Janet, who was then with us, and afraid even to speak of death, was horrified on seeing them, firmly believing that she would one day be eaten by them—a very general opinion at that time; few people being then aware that the finest mould in our gardens and fields has passed through the entrails of the earthworm, the vegetable juices it contains being sufficient to maintain these harmless creatures.

My mother was very much afraid of thunder and lightning. She knew when a storm was near from the appearance of the clouds, and prepared for it by taking out the steel pins which fastened her cap on. She then sat on a sofa at a distance from the fireplace, which had a very high chimney, and read different parts of the Bible, especially the sublime descriptions of storms in the Psalms, which made me, who sat close by her, still more afraid. We had an excellent and beautiful pointer, called Hero, a great favourite, who generally lived in the garden, but at the first clap of thunder he used to rush howling in-doors, and place his face on my knee. Then my father, who laughed not a little at our fear, would bring a glass of wine to my mother, and say, "Drink that, Peg; it will give you courage, for we are going to have a rat-tat-too." My mother would beg him to shut the window-shutters, and though she could no longer see to read, she kept the Bible on her knee for protection.

My mother taught me to read the Bible, and to say my prayers morning and evening; otherwise she allowed me to grow up a wild creature. When I was seven or eight years old I began to be useful, for I pulled the fruit for preserving; shelled the peas and beans, fed the poultry, and looked after the dairy, for we kept a cow.

On one occasion I had put green gooseberries into bottles and sent them to the kitchen with orders to the cook to boil the bottles uncorked, and, when the fruit was sufficiently cooked, to cork and tie up the bottles. After a time all the house was alarmed by loud explosions and violent screaming in the kitchen; the cook had corked the bottles before she boiled them, and of course they exploded. For greater preservation, the bottles were always buried in the ground; a number were once found in our garden with the fruit in high preservation which had been buried no one knew when. Thus experience is sometimes the antecedent of science, for it was little suspected at that time that by shutting out the air the invisible organic world was excluded—the cause of all fermentation and decay.

I never cared for dolls, and had no one to play with me. I amused myself in the garden, which was much frequented by birds. I knew most of them, their flight and their habits. The swallows were never prevented from building above our windows, and, when about to migrate, they used to assemble in hundreds on the roof of our house, and prepared for their journey by short flights. We fed the birds when the ground was covered with snow, and opened our windows at breakfast-time to let in the robins, who would hop on the table to pick up crumbs. The quantity of singing birds was very great, for the farmers and gardeners were less cruel and avaricious than they are now—though poorer. They allowed our pretty songsters to share in the bounties of providence. The shortsighted cruelty, which is too prevalent now, brings its own punishment, for, owing to the reckless destruction of birds, the equilibrium of nature is disturbed, insects increase to such an extent as materially to affect every description of crop. This summer (1872), when I was at Sorrento, even the olives, grapes, and oranges were seriously injured by the caterpillars—a disaster which I entirely attribute to the ruthless havoc made among every kind of bird.

* * * * *

My mother set me in due time to learn the catechism of the Kirk of Scotland, and to attend the public examinations in the kirk. This was a severe trial for me ; for, besides being timid and shy, I had a bad memory, and did not understand one word of the catechism. These meetings, which began with prayer, were attended by all the children of the town and neighbourhood, with their mothers, and a great many old women, who came to be edified. They were an acute race, and could quote chapter and verse of Scripture as accurately as the minister himself. I remember he said to one of them—" Peggie, what lightened the world before the sun was made ? " After thinking for a minute, she said—" 'Deed, sir, the question is mair curious than edifying."

Besides these public examinations, the minister made an annual visit to each household in his parish. When he came to us, the servants were called in, and we all knelt while he said a prayer ; and then he examined each individual as to the state of his soul and conduct. He asked me if I could say my "Questions"—that is, the catechism of the Kirk of Scotland—and asked a question at random to ascertain the fact. He did the same to the servants.

When I was between eight and nine years old, my father came home from sea, and was shocked to find me such a savage. I had not yet been taught to write, and although I amused myself reading the "Arabian Nights," "Robinson Crusoe," and the "Pilgrim's Progress," I read very badly, and with a strong Scotch accent; so, besides a chapter of the Bible, he made me read a paper of the "Spectator" aloud every morning, after breakfast; the consequence of which discipline is that I have never since opened that book. Hume's "History of England" was also a real penance to me. I gladly accompanied my father when he cultivated his flowers, which even now I can say were of the best quality. The tulips and other bulbous plants, ranunculi, anemones, carnations, as well as the annuals then known, were all beautiful. He used to root up and throw away many plants I thought very beautiful; he said he did so because the colours of their petals were not sharply defined, and that they would spoil the seed of the others. Thus I learnt to know the good and the bad—how to lay carnations, and how to distinguish between the leaf and fruit buds in pruning fruit trees; this kind of knowledge was of no practical use, for, as my after-life was spent in towns, I never had a garden, to my great regret.

George the Third was so popular, that even in Burntisland nosegays were placed in every window on the 4th of June, his birthday; and it occasionally happened that our garden was robbed the preceding night of its gayest flowers.

My father at last said to my mother,—"This kind of life will never do, Mary must at least know how to write and keep accounts." So at ten years old I was sent to a boarding-school, kept by a Miss Primrose, at Musselburgh, where I was utterly wretched. The change from perfect liberty to perpetual restraint was in itself a great trial; besides, being naturally shy and timid, I was afraid of strangers, and although Miss Primrose was not unkind she had an habitual frown, which even the elder girls dreaded. My future companions, who were all older than I, came round me like a swarm of bees, and asked if my father had a title, what was the name of our estate, if we kept a carriage, and other such questions, which made me first feel the difference of station. However, the girls were very kind, and often bathed my eyes to prevent our stern mistress from seeing that I was perpetually in tears. A few days after my arrival, although perfectly straight and well-made, I was enclosed in stiff stays with a steel busk in front, while, above my frock, bands drew my shoulders back till the shoulder-blades met. Then a steel rod, with a semi-circle which went under the chin, was clasped to the steel busk in my stays. In this constrained state I, and most of the younger girls, had to prepare our lessons. The chief thing I had to do was to learn by heart a page of Johnson's dictionary, not only to spell the words, give their parts of speech and meaning, but as an exercise of memory to remember their order of succession. Besides I had to learn the first principles of writing, and the rudiments of French and English grammar. The method of teaching was extremely tedious and inefficient. Our religious duties were attended to in a remarkable way. Some of the girls were Presbyterians, others belonged to the Church of England, so Miss Primrose cut the matter short by taking us all to the kirk in the morning and to church in the afternoon.

In our play-hours we amused ourselves with playing at ball, marbles, and especially at "Scotch and English," a game which represented a raid on the debatable land, or Border between Scotland and England, in which each party tried to rob the other of their playthings. The little ones were always compelled to be English, for the bigger girls thought it too degrading.

Lady Hope, a relative of my mother, frequently invited me to spend Saturday at Pinkie. She was a very ladylike person, in delicate health, and with cold manners. Sir Archibald was stout, loud, passionate, and devoted to hunting. I amused myself in the grounds, a good deal afraid of a turkey-cock, who was pugnacious and defiant.

CHAPTER II.

FREEDOM—RELIGIOUS EDUCATION—JEDBURGH.

[My mother remained at school at Musselburgh for a twelvemonth, till she was eleven years old. After this prolonged and elaborate education, she was recalled to Burntisland, and the results of the process she had undergone are detailed in her "Recollections" with much drollery.

SOON after my return home I received a note from a lady in the neighbourhood, inquiring for my mother, who had been ill. This note greatly distressed me, for my half-text writing was as bad as possible, and I could neither compose an answer nor spell the words. My eldest cousin, Miss Somerville, a grown-up young lady, then with us, got me out of this scrape, but I soon got myself into another, by writing to my brother in Edinburgh that I had sent him a bank-*knot* (note) to buy something for me. The school at Musselburgh was expensive, and I was reproached with having cost so much money in vain. My mother said she would have been contented if I had only learnt to write well and keep accounts, which was all that a woman was expected to know.

This passed over, and I was like a wild animal escaped out of a cage. I was no longer amused in the gardens, but wandered about the country. When the tide was out I spent hours on the sands, looking at the star-fish and sea-urchins, or watching the children digging for sand-eels, cockles, and the spouting razor-fish. I made a collection of shells, such as were cast ashore, some so small that they appeared like white specks in patches of black sand. There was a small pier on the sands for shipping limestone brought from the coal mines inland. I was astonished to see the surface of these blocks of stone covered with beautiful impressions of what seemed to be leaves; how they got there I could not imagine, but I picked up the broken bits, and even large pieces, and brought them to my repository. I knew the eggs of many birds, and made a collection of them. I never robbed a nest, but bought strings of eggs, which were sold by boys, besides getting sea-fowl eggs from sailors who had been in whalers or on other northern voyages. It was believed by these sailors that there was a gigantic flat fish in the North Sea, called a kraken. It was so enormous that when it came to the surface, covered with tangles and sand, it was supposed to be an island, till, on one occasion, part of a ship's crew landed on it and found out their mistake. However, much as they believed in it, none of the sailors at Burntisland had ever seen it. The sea serpent was also an article of our faith.

In the rocks at the end of our garden there was a shingly opening, in which we used to bathe, and where at low tide I frequently waded among masses of rock covered with sea-weeds. With the exception of dulse and tangle I knew the names of none, though I was well acquainted with and admired many of these beautiful plants. I also watched the crabs, live shells, jelly-fish, and various marine animals, all of which were objects of curiosity and amusement to me in my lonely life.

The flora on the links and hills around was very beautiful, and I soon learnt the trivial names of all the plants. There was not a tree nor bush higher than furze in this part of the country, but the coast to the north-west of Burntisland was bordered by a tree and brushwood-covered bank belonging to the Earl of Morton, which extended to Aberdour. I could not go so far alone, but had frequent opportunities of walking there and gathering ferns, foxgloves, and primroses, which grew on the mossy banks of a little stream that ran into the sea. The bed of this stream or burn was thickly covered with the freshwater mussel, which I knew often contained pearls, but I did not like to kill the creatures to get the pearls.

One day my father, who was a keen sportsman, having gone to fish for red trout at the mouth of this stream, found a young whale, or grampus, stranded in the shallow water. He immediately ran back to the town, got boats, captured the whale, and landed it in the harbour, where I went with the rest of the crowd to see the *muckle fish.*

There was always a good deal of shipbuilding carried on in the harbour, generally coasting vessels or colliers. We, of course, went to see them launched, which was a pretty sight.

* * * * *

When the bad weather began I did not know what to do with myself. Fortunately we had a small collection of books, among which I found Shakespeare, and read it at every moment I could spare from my domestic duties. These occupied a great part of my time; besides, I had to *shew* (sew) my sampler, working the alphabet from A to Z, as well as the ten numbers, on canvas.

My mother did not prevent me from reading, but my aunt Janet, who came to live in Burntisland after her father's death, greatly disapproved of my conduct. She was an old maid who could be very agreeable and witty, but she had all the prejudices of the time with regard to women's duties, and said to my mother, "I wonder you let Mary waste her time in reading, she never *shews* (sews) more than if she were a man." Whereupon I was sent to the village school to learn plain needlework. I do not remember how long it was after this that an old lady sent some very fine linen to be made into shirts for her brother, and desired that one should be made entirely by me. This shirt was so well worked that I was relieved from attending the school, but the house linen was given into my charge to make and to mend. We had a large stock, much of it very beautiful, for the Scotch ladies at that time were very proud of their napery, but they no longer sent it to Holland to be bleached, as had once been the custom. We grew flax, and our maids spun it. The coarser yarn was woven in Burntisland, and bleached upon the links; the finer was sent to Dunfermline, where there was a manufactory of table-linen.

I was annoyed that my turn for reading was so much disapproved of, and thought it unjust that women should have been given a desire for knowledge if it were wrong to acquire it. Among our books I found Chapone's "Letters to Young Women," and resolved to follow the course of history there recommended, the more so as we had most of the works she mentions. One, however, which my cousin lent me was in French, and here the little I had learnt at school was useful, for with the help of a dictionary I made out the sense. What annoyed me was my memory not being good—I could remember neither names nor dates. Years afterwards I studied a "Memoria Technica," then in fashion, without success; yet in my youth I could play long pieces of music on the piano without the book, and I never forget mathematical formulæ. In looking over one of my MSS., which I had not seen for forty years, I at once recognised the formulæ for computing the secular inequalities of the moon.

We had two small globes, and my mother allowed me to learn the use of them from Mr. Reed, the village schoolmaster, who came to teach me for a few weeks in the winter evenings. Besides the ordinary branches, Mr. Reed taught Latin and navigation, but these were out of the question for me. At the village school the boys often learnt Latin, but it was thought sufficient for the girls to be able to read the Bible; very few even learnt writing. I recollect, however, that some men were ignorant of book-keeping; our baker, for instance, had a wooden tally, in which he made a notch for every loaf of bread, and of course we had the corresponding tally. They were called nick-sticks.

My bedroom had a window to the south, and a small closet near had one to the north. At these I spent many hours, studying the stars by the aid of the celestial globe. Although I watched and admired the magnificent displays of the Aurora, which frequently occurred, they seemed to be so nearly allied to lightning that I was somewhat afraid of them. At an earlier period of my life there was a comet, which I dreaded exceedingly.

* * * * *

My father was Captain of the "Repulse," a fifty-gun ship, attached to the Northern fleet commanded by the Earl of Northesk. The winter was extremely stormy, the fleet was driven far north, and kept there by adverse gales, till both officers and crew were on short rations. They ran out of candles, and had to tear up their stockings for wicks, and dip them into the fat of the salt meat which was left. We were in great anxiety, for it was reported that some of the ships had foundered; we were, however, relieved by the arrival of the "Repulse" in Leith roads for repair.

Our house on one occasion being full, I was sent to sleep in a room quite detached from the rest and with a different staircase. There was a closet in this room in which my father kept his fowling pieces, fishing tackle, and golf clubs, and a long garret overhead was filled with presses and stores of all kinds, among other things a number of large cheeses were on a board slung by ropes to the rafters. One night I had put out my candle and was fast asleep, when I was awakened by a violent crash, and then a rolling noise over my head. Now the room was said to be haunted, so that the servants would not sleep in it. I was desperate, for there was no bell. I groped my way to the closet—lucifer matches were unknown in those days—I seized one of the golf clubs, which are shod with iron, and thundered on the bedroom door till I brought my father, followed by the whole household, to my aid. It was found that the rats had gnawed through the ropes by which the cheeses were suspended, so that the crash and rolling were accounted for, and I was scolded for making such an uproar.

Children suffer much misery by being left alone in the dark. When I was very young I was sent to bed at eight or nine o'clock, and the maid who slept in the room went away as soon as I was in bed, leaving me alone in the dark till she came to bed herself. All that time I was in an agony of fear of something indefinite, I could not tell what. The joy, the relief, when the maid came back, were such that I instantly fell asleep. Now that I am a widow and old, although I always have a night-lamp, such is the power of early impressions that I rejoice when daylight comes.

* * * * *

At Burntisland the sacrament was administered in summer because people came in crowds from the neighbouring parishes to attend the preachings. The service was long and fatiguing. A number of clergymen came to assist, and as the minister's manse could not accommodate them all, we entertained three of them, one of whom was always the Rev. Dr. Campbell, father of Lord Campbell.

Thursday was a day of preparation. The morning service began by a psalm sung by the congregation, then a prayer was said by the minister, followed by a lecture on some chapter of the Bible, generally lasting an hour, after that another psalm was sung, followed by a prayer, a sermon which lasted seldom less than an hour, and the whole ended with a psalm, a short prayer and a benediction. Every one then went home to dinner and returned afterwards for afternoon service, which lasted more than an hour and a half. Friday was a day of rest, but I together with many young people went at this time to the minister to receive a stamped piece of lead as a token that we were sufficiently instructed to be admitted to Christ's table. This ticket was given to the Elder on the following Sunday. On Saturday there was a morning service, and on Sunday such multitudes came to receive the sacrament that the devotions continued till late in the evening. The ceremony was very strikingly and solemnly conducted. The communicants sat on each side of long narrow tables covered with white linen, in imitation of the last supper of Christ, and the Elders handed the bread and wine. After a short exhortation from one of the ministers the first set retired, and were succeeded by others. When the weather was fine a sermon, prayers, and psalm-singing took place either in the churchyard or on a grassy bank at the Links for such as were waiting to communicate. On the Monday morning there was the same long service as on the Thursday. It was too much for me; I always came home with a headache, and took a dislike to sermons.

Our minister was a rigid Calvinist. His sermons were gloomy, and so long that he occasionally would startle the congregation by calling out to some culprit, "Sit up there, how daur ye sleep i' the kirk." Some saw-mills in the neighbourhood were burnt down, so the following Sunday we had a sermon on hell-fire. The kirk was very large and quaint; a stair led to a gallery on each side of the pulpit, which was intended for the tradespeople, and each division was marked with a suitable device, and text from Scripture. On the bakers' portion a sheaf of wheat was painted; a balance and weights on the grocers', and on the weavers', which was opposite to our pew, there was a shuttle, and below it the motto, "My days are swifter than a weaver's shuttle, and are spent without *hop job.*" The artist was evidently no clerk.

My brother Sam, while attending the university in Edinburgh, came to us on the Saturdays and returned to town on Monday. He of course went with us to the kirk on Sunday morning, but we let our mother attend afternoon service alone, as he and I were happy to be together, and we spent the time sitting on the grassy rocks at the foot of our garden, from whence we could see a vast extent of the Firth of Forth with Edinburgh and its picturesque hills. It was very amusing, for we occasionally saw three or four whales spouting, and shoals of porpoises at play. However, we did not escape reproof, for I recollect the servant coming to tell us that the minister had sent to inquire whether Mr. and Miss Fairfax had been taken ill, as he had not seen them at the kirk in the afternoon. The minister in question was Mr. Wemyss, who had married a younger sister of my mother's.

* * * * *

When I was about thirteen my mother took a small apartment in Edinburgh for the winter, and I was sent to a writing school, where I soon learnt to write a good hand, and studied the common rules of arithmetic. My uncle William Henry Charters, lately returned from India, gave me a pianoforte, and I had music lessons from an old lady who lived in the top story of one of the highest houses in the old town. I slept in the same room with my mother. One morning I called out, much alarmed, "There is lightning!" but my mother said, after a moment, "No; it is fire!" and on opening the window shutters I found that the flakes of fire flying past had made the glass quite hot. The next house but one was on fire and burning fiercely, and the people next door were throwing everything they possessed, even china and glass, out of the windows into the street. We dressed quickly, and my mother sent immediately to Trotter the upholsterer for four men. We then put our family papers, our silver, &c., &c., into trunks; then my mother said, "Now let us breakfast, it is time enough for us to move our things when the next house takes fire." Of its doing so there was every probability because casks of turpentine and oil were exploding from time to time in a carriage manufactory at the back of it. Several gentlemen of our acquaintance who came to assist us were surprised to find us breakfasting quietly as if there were nothing unusual going on. In fact my mother, though a coward in many things, had, like most women, the presence of mind and the courage of necessity. The fire was extinguished, and we had only the four men to pay for doing nothing, nor did we sacrifice any of our property like our neighbours who had completely lost their heads from terror. I may mention here that on one occasion when my father was at home he had been ill with a severe cold, and wore his nightcap. While reading in the drawing-room one evening he called out, "I smell fire, there is no time to be lost," so, snatching up a candle, he wandered from room to room followed by us all still smelling fire, when one of the servants said, "O, sir, it is the tassel of your nightcap that is on fire."

* * * * *

On returning to Burntisland, I spent four or five hours daily at the piano; and for the sake of having something to do, I taught myself Latin enough, from such books as we had, to read Cæsar's "Commentaries." I went that summer on a visit to my aunt at Jedburgh, and, for the first time in my life, I met in my uncle, Dr. Somerville, with a friend who approved of my thirst for knowledge. During long walks with him in the early mornings, he was so kind, that I had the courage to tell him that I had been trying to learn Latin, but I feared it was in vain; for my brother and other boys, superior to me in talent, and with every assistance, spent years in learning it. He assured me, on the contrary, that in ancient times many women—some of them of the highest rank in England—had been very elegant scholars, and that he would read Virgil with me if I would come to his study for an hour or two every morning before breakfast, which I gladly did.

I never was happier in my life than during the months I spent at Jedburgh. My aunt was a charming companion—witty, full of anecdote, and had read more than most women of her day, especially Shakespeare, who was her favourite author. My cousins had little turn for reading, but they were better educated than most girls. They were taught to write by David Brewster, son of the village schoolmaster, afterwards Sir David, who became one of the most distinguished philosophers and discoverers of the age, member of all the scientific societies at home and abroad, and at last President of the University of Edinburgh. He was studying in Edinburgh when I was at Jedburgh; so I did not make his acquaintance then; but later in life he became my valued friend. I did not know till after his death, that, while teaching my cousins, he fell in love with my cousin Margaret. I do not believe she was aware of it. She was afterwards attached to an officer in the army; but my aunt would not allow her to go to that *outlandish* place, Malta, where he was quartered; so she lived and died unmarried. Steam has changed our ideas of distance since that time.

My uncle's house—the manse—in which I was born, stands in a pretty garden, bounded by the fine ancient abbey, which, though partially ruined, still serves as the parish kirk. The garden produced abundance of common flowers, vegetables, and fruit. Some of the plum and pear trees were very old, and were said to have been planted by the monks. Both were excellent in quality, and very productive. The view from both garden and manse was over the beautiful narrow valley through which the Jed flows. The precipitous banks of red sandstone are richly clothed with vegetation, some of the trees ancient and very fine, especially the magnificent one called the capon tree, and the lofty king of the wood, remnants of the fine forests which at one time had covered the country. An inland scene was new to me, and I was never tired of admiring the tree-crowned scaurs or precipices, where the rich glow of the red sandstone harmonized so well with the autumnal tints of the foliage.

We often bathed in the pure stream of the Jed. My aunt always went with us, and was the merriest of the party; we bathed in a pool which was deep under the high scaur, but sloped gradually from the grassy bank on the other side. Quiet and transparent as the Jed was, it one day came down with irresistible fury, red with the débris of the sandstone scaurs. There had been a thunderstorm in the hills up-stream, and as soon as the river began to rise, the people came out with pitchforks and hooks to catch the hayricks, sheaves of corn, drowned pigs, and other animals that came sweeping past. My cousins and I were standing on the bridge, but my aunt called us off when the water rose above the arches, for fear of the bridge giving way. We made expeditions every day; sometimes we went nutting in the forest; at other times we gathered mushrooms on the grass parks of Stewartfield, where there was a wood of picturesque old Scotch firs, inhabited by a colony of rooks. I still kept the habit of looking out for birds, and had the good fortune to see a heron, now a rare bird in the valley of the Jed. Some of us went every day to a spring called the Allerly well, about a quarter of a mile from the manse, and brought a large jug of its sparkling water for dinner. The evenings were cheerful; my aunt sang Scotch songs prettily, and told us stories and legends about Jedburgh, which had been a royal residence in the olden time. She had a tame white and tawny-coloured owl, which we fed every night, and sometimes brought into the drawing-room. The Sunday evening never was gloomy, though properly observed. We occasionally drank tea with acquaintances, and made visits of a few days to the Rutherfurds of Edgerton and others; but I was always glad to return to the manse.

My uncle, like other ministers of the Scottish Kirk, was allowed a glebe, which he farmed himself. Besides horses, a cow was kept, which supplied the family with cream and butter, and the skimmed milk was given to the poor; but as the milk became scarce, one woman was deprived, for the time, of her share. Soon after, the cow was taken ill, and my uncle's ploughman, Will, came to him and said, "Sir, gin you would give that carline Tibby Jones her soup o' milk again, the coo would soon be weel eneugh." Will was by no means the only believer in witchcraft at that time.

CHAPTER III.

EDINBURGH—YOUTHFUL STUDIES AND AMUSEMENTS—POLITICS—THE THEATRES OF THE TIME.

[My mother's next visit was to the house of her uncle, William Charters, in Edinburgh. From thence she was enabled to partake of the advantages of a dancing-school of the period.

THEY sent me to Strange's dancing school. Strange himself was exactly like a figure on the stage; tall and thin, he wore a powdered wig, with cannons at the ears, and a pigtail. Ruffles at the breast and wrists, white waistcoat, black silk or velvet shorts, white silk stockings, large silver buckles, and a pale blue coat completed his costume. He had a little fiddle on which he played, called a kit. My first lesson was how to walk and make a curtsey. "Young lady, if you visit the queen you must make three curtsies, lower and lower and lower as you approach her. So—o—o," leading me on and making me curtsey. "Now, if the queen were to ask you to eat a bit of mutton with her, what would you say?"

Every Saturday afternoon all the scholars, both boys and girls, met to practise in the public assembly rooms in George's Street. It was a handsome large hall with benches rising like an amphitheatre. Some of the elder girls were very pretty, and danced well, so these practisings became a lounge for officers from the Castle, and other young men. We used always to go in full evening dress. We learnt the *minuet de la cour*, reels and country dances. Our partners used to give us gingerbread and oranges. Dancing before so many people was quite an exhibition, and I was greatly mortified one day when ready to begin a minuet, by the dancing-master shaking me roughly and making me hold out my frock properly.

Though kind in the main, my uncle and his wife were rather sarcastic and severe, and kept me down a good deal, which I felt keenly, but said nothing. I was not a favourite with my family at that period of my life, because I was reserved and unexpansive, in consequence of the silence I was obliged to observe on the subjects which interested me. Three Miss Melvilles, friends, or perhaps relatives, of Mrs. Charters, were always held up to me as models of perfection, to be imitated in everything, and I wearied of hearing them constantly praised at my expense.

In a small society like that of Edinburgh there was a good deal of scandal and gossip; every one's character and conduct were freely criticised, and by none more than by my aunt and her friends. She used to sit at a window embroidering, where she not only could see every one that passed, but with a small telescope could look into the dressing-room of a lady of her acquaintance, and watch all she did. A spinster lady of good family, a cousin of ours, carried her gossip so far, that she was tried for defamation, and condemned to a month's imprisonment, which she actually underwent in the Tolbooth. She was let out just before the king's birthday, to celebrate which, besides the guns fired at the Castle, the boys let off squibs and crackers in all the streets. As the lady in question was walking up the High Street, some lads in a wynd, or narrow street, fired a small cannon, and one of the slugs with which it was loaded hit her mouth and wounded her tongue. This raised a universal laugh; and no one enjoyed it more than my uncle William, who disliked this somewhat masculine woman.

Whilst at my uncle's house, I attended a school for writing and arithmetic, and made considerable progress in the latter, for I liked it, but I soon forgot it from want of practice.

My uncle and aunt generally paid a visit to the

Lyells of Kinnordy, the father and mother of my friend Sir Charles Lyell, the celebrated geologist; but this time they accepted an invitation from Captain Wedderburn, and took me with them. Captain Wedderburn was an old bachelor, who had left the army and devoted himself to agriculture. Mounted on a very tall but quiet horse, I accompanied my host every morning when he went over his farm, which was chiefly a grass farm. The house was infested with rats, and a masculine old maid, who was of the party, lived in such terror of them, that she had a light in her bedroom, and after she was in bed, made her maid tuck in the white dimity curtains all round. One night we were awakened by violent screams, and on going to see what was the matter, we found Miss Cowe in the middle of the room, bare-footed, in her night-dress, screaming at the top of her voice. Instead of tucking the rats out of the bed, the maid had tucked one in, and Miss Cowe on waking beheld it sitting on her pillow.

* * * * *

There was great political agitation at this time. The corruption and tyranny of the court, nobility, and clergy in France were so great, that when the revolution broke out, a large portion of our population thought the French people were perfectly justified in revolting, and warmly espoused their cause. Later many changed their opinions, shocked, as every one was, at the death of the king and queen, and the atrocious massacres which took place in France. Yet some not only approved of the revolution abroad, but were so disgusted with our mal-administration at home, to which they attributed our failure in the war in Holland and elsewhere, that great dissatisfaction and alarm prevailed throughout the country. The violence, on the other hand, of the opposite party was not to be described,—the very name of Liberal was detested.

Great dissensions were caused by difference of opinion in families; and I heard people previously much esteemed accused from this cause of all that was evil. My uncle William and my father were as violent Tories as any.

The Liberals were distinguished by wearing their hair short, and when one day I happened to say how becoming a crop was, and that I wished the men would cut off those ugly pigtails, my father exclaimed, "By G—, when a man cuts off his queue, the head should go with it."

The unjust and exaggerated abuse of the Liberal party made me a Liberal. From my earliest years my mind revolted against oppression and tyranny, and I resented the injustice of the world in denying all those privileges of education to my sex which were so lavishly bestowed on men. My liberal opinions, both in religion and politics, have remained unchanged (or, rather, have advanced) throughout my life, but I have never been a republican. I have always considered a highly-educated aristocracy essential, not only for government, but for the refinement of a people.

[After her winter in Edinburgh, my mother returned to Burntisland. Strange to say, she found there, in an illustrated Magazine of Fashions, the introduction to the great study of her life.

I was often invited with my mother to the tea-parties given either by widows or maiden ladies who resided at Burntisland. A pool of commerce used to be keenly contested till a late hour at these parties, which bored me exceedingly, but I there became acquainted with a Miss Ogilvie, much younger than the rest, who asked me to go and see fancy works she was doing, and at which she was very clever. I went next day, and after admiring her work, and being told how it was done, she showed me a monthly magazine with coloured plates of ladies' dresses, charades, and puzzles. At the end of a page I read what appeared to me to be simply an arithmetical question ; but on turning the page I was surprised to see strange looking lines mixed with letters, chiefly X'es and Y's, and asked; "What is that?" "Oh," said Miss Ogilvie, "it is a kind of arithmetic: they call it Algebra; but I can tell you nothing about it." And we talked about other things; but on going home I thought I would look if any of our books could tell me what was meant by Algebra.

In Robertson's "Navigation" I flattered myself that I had got precisely what I wanted; but I soon found that I was mistaken. I perceived, however, that astronomy did not consist in star-gazing,* and as I persevered in studying the book for a time, I certainly got a dim view of several subjects which were useful to me afterwards. Unfortunately not one of our acquaintances or relations knew anything of science or natural history; nor, had they done so, should I have had courage to ask any of them a question, for I should have been laughed at. I was often very sad and forlorn; not a hand held out to help me.

My uncle and aunt Charters took a house at Burntisland for the summer, and the Miss Melville I have already mentioned came to pay them a visit. She

* Many people evidently think the science of astronomy consists entirely in observing the stars, for I have been frequently asked if I passed my nights looking through a telescope, and I have astonished the enquirers by saying I did not even possess one.

painted miniatures, and from seeing her at work, I took a fancy to learn to draw, and actually wasted time in copying prints; but this circumstance enabled me to get elementary books on Algebra and Geometry without asking questions of any one, as will be explained afterwards. The rest of the summer I spent in playing on the piano and learning Greek enough to read Xenophon and part of Herodotus; then we prepared to go to Edinburgh.

My mother was so much afraid of the sea that she never would cross the Firth except in a boat belonging to a certain skipper who had served in the Navy and lost a hand; he had a hook fastened on the stump to enable him to haul ropes. My brother and I were tired of the country, and one sunny day we persuaded my mother to embark. When we came to the shore, the skipper said, "I wonder that the leddy boats to-day, for though it is calm here under the lee of the land, there is a stiff breeze outside." We made him a sign to hold his tongue, for we knew this as well as he did. Our mother went down to the cabin and remained silent and quiet for a time; but when we began to roll and be tossed about, she called out to the skipper, "George! this is an awful storm, I am sure we are in great danger. Mind how you steer; remember, I trust in you!" He laughed, and said, "Dinna trust in me, leddy;

trust in God Almighty." Our mother, in perfect terror, called out, "Dear me! is it come to that?" We burst out laughing, skipper and all.

Nasmyth, an exceedingly good landscape painter, had opened an academy for ladies in Edinburgh, a proof of the gradual improvement which was taking place in the education of the higher classes; my mother, very willingly allowed me to attend it. The class was very full. I was not taught to draw, but looked on while Nasmyth painted; then a picture was given me to copy, the master correcting the faults. Though I spoilt canvas, I had made some progress by the end of the season.* Mr. Nasmyth, besides being a good artist, was clever, well-informed, and had a great deal of conversation. One day I happened to be near him while he was talking to the Ladies Douglas about perspective. He said, "You should study Euclid's Elements of Geometry, the foundation not only of perspective, but of astronomy and all mechanical science." Here, in the most unexpected manner, I got the information I wanted, for I at once saw that it would help me to understand some parts of Robertson's "Navigation;" but as to going to a bookseller and asking for Euclid the

* Nasmyth told a lady still alive who took lessons from him in her youth, that the cleverest young lady he ever taught was Miss Mary Fairfax.

thing was impossible! Besides I did not yet know anything definite about Algebra, so no more could be done at that time; but I never lost sight of an object which had interested me from the first.

I rose early, and played four or five hours, as usual, on the piano, and had lessons from Corri, an Italian, who taught carelessly, and did not correct a habit I had of thumping so as to break the strings; but I learned to tune a piano and mend the strings, as there was no tuner at Burntisland. Afterwards I got over my bad habit and played the music then in vogue: pieces by Pleyel, Clementi, Steibelt, Mozart, and Beethoven, the last being my favourite to this day. I was sometimes accompanied on the violin by Mr. Thomson, the friend of Burns; more frequently by Stabilini; but I was always too shy to play before people, and invariably played badly when obliged to do so, which vexed me.

* * * * *

The prejudice against the theatre had been very great in Scotland, and still existed among the rigid Calvinists. One day, when I was fourteen or fifteen, on going into the drawing-room, an old man sitting beside my mother rose and kissed me, saying, "I am one of your mother's oldest friends." It was Home, the author of the tragedy of "Douglas." He was obliged to resign his living in the kirk for the scandal of having had his play acted in the theatre in Edinburgh, and some of his clerical friends were publicly rebuked for going to see it. Our family was perfectly liberal in all these matters. The first time I had ever been in a theatre I went with my father to see "Cymbeline." I had never neglected Shakespeare, and when our great tragedians, Mrs. Siddons and her brother, John Kemble, came for a short time to act in Edinburgh, I could think of nothing else. They were both remarkably handsome, and, notwithstanding the Scotch prejudice, the theatre was crowded every night. It was a misfortune to me that my mother never would go into society during the absence of my father, nor, indeed, at any time, except, perhaps, to a dinner party; but I had no difficulty in finding a chaperone, as we knew many people. I used to go to the theatre in the morning, and ask to see the plan of the house for the evening, that I might know which ladies I could accompany to their boxes. Of course I paid for my place. Our friends were so kind that I saw these great artists, as well as Charles Kemble, Young, and Bannister, in "Hamlet," "Macbeth," "Othello," "Coriolanus," "The Gamester," &c.

It was greatly to the honour of the British stage that all the principal actors, men and women, were of excellent moral character, and much esteemed. Many years afterwards, when Mrs. Siddons was an old woman, I drank tea with her, and heard her read Milton and Shakespeare. Her daughter told us to applaud, for she had been so much accustomed to it in the theatre that she could not read with spirit without this expression of approbation.

My mother was pleased with my music and painting, and, although she did not go to the theatre herself, she encouraged me to go. She was quite of the old school with regard to the duties of women, and very particular about her table; and, although we were obliged to live with rigid economy, our food was of the best quality, well dressed, and neatly served, for she could tell the cook exactly what was amiss when anything was badly cooked. She thought besides that some of the comfort of married life depended upon the table, so I was sent to a pastrycook for a short time every day, to learn the art of cookery. I had for companions Miss Moncreiff, daughter of Sir Henry Moncreiff Wellwood, a Scotch baronet of old family. She was older than I, pretty, pleasing, and one of the belles of the day. We were amused at the time, and afterwards made jellies and creams for little supper parties, then in fashion, though, as far as economy went, we might as well have bought them.

On returning to Burntisland, I played on the piano as diligently as ever, and painted several hours every day. At this time, however, a Mr. Craw came to live with us as tutor to my youngest brother, Henry. He had been educated for the kirk, was a fair Greek and Latin scholar, but, unfortunately for me, was no mathematician. He was a simple, good-natured kind of man, and I ventured to ask him about algebra and geometry, and begged him, the first time he went to Edinburgh, to buy me something elementary on these subjects, so he soon brought me "Euclid" and Bonnycastle's "Algebra," which were the books used in the schools at that time. Now I had got what I so long and earnestly desired. I asked Mr. Craw to hear me demonstrate a few problems in the first book of "Euclid," and then I continued the study alone with courage and assiduity, knowing I was on the right road. Before I began to read algebra I found it necessary to study arithmetic again, having forgotten much of it. I never was expert at addition, for, in summing up a long column of pounds, shillings, and pence, in the family account book, it seldom came out twice the same way. In after life I, of course, used logarithms for the higher branches of science.

I had to take part in the household affairs, and to

make and mend my own clothes. I rose early, played on the piano, and painted during the time I could spare in the daylight hours, but I sat up very late reading Euclid. The servants, however, told my mother "It was no wonder the stock of candles was soon exhausted, for Miss Mary sat up reading till a very late hour;" whereupon an order was given to take away my candle as soon as I was in bed. I had, however, already gone through the first six books of Euclid, and now I was thrown on my memory, which I exercised by beginning at the first book, and demonstrating in my mind a certain number of problems every night, till I could nearly go through the whole. My father came home for a short time, and, somehow or other, finding out what I was about, said to my mother, "Peg, we must put a stop to this, or we shall have Mary in a strait jacket one of these days. There was X., who went raving mad about the longitude!"

* * * * *

In our younger days my brother Sam and I kept various festivals: we burnt nuts, ducked for apples, and observed many other of the ceremonies of Halloween, so well described by Burns, and we always sat up to hail the new year on New Year's Eve. When in Edinburgh we sometimes disguised ourselves as "guisarts," and went about with a basket full of Christmas cakes called buns and shortbread, and a flagon of "het-pint" or posset, to wish our friends a "Happy New Year." At Christmas time a set of men, called the Christmas Wakes, walked slowly through the streets during the midnight hours, playing our sweet Scotch airs on flageolets. I remember the sound from a distance fell gently on my sleeping ear, swelled softly, and died away in distance again, a passing breeze of sweet sound. It was very pleasing; some thought it too sad.

My grandfather was intimate with the Boswells of Balmuto, a bleak place a few miles to the north of Burntisland. Lord Balmuto, a Scotch judge, who was then proprietor, had been a dancing companion of my mother's, and had a son and two daughters, the eldest a nice girl of my age, with whom I was intimate, so I gladly accepted an invitation to visit them at Balmuto. Lord Balmuto was a large coarse-looking man, with black hair and beetling eyebrows. Though not vulgar, he was passionate, and had a boisterous manner. My mother and her sisters gave him the nickname of the "black bull of Norr'away," in allusion to the northern position of Balmuto. Mrs. Boswell was gentle and lady-like. The son had a turn for chemistry, and his father took me to see what they called the Laboratory. What a laboratory might be I knew not, as I had never heard the word before, but somehow I did not like the look of the curiously-shaped glass things and other apparatus, so when the son put a substance on the table, and took a hammer, his father saying, "Now you will hear a fine report," I ran out of the room, saying, "I don't like reports." Sure enough there was a very loud report, followed by a violent crash, and on going into the room again, we found that the son had been knocked down, the father was trembling from head to foot, and the apparatus had been smashed to pieces. They had had a narrow escape. Miss Boswell led a dull life, often passing the winter with her mother in that solitary place, Balmuto; and when in Edinburgh, she was much kept down by her father, and associated little with people of her own age and station. The consequence was that she eloped with her drawing-master, to the inexpressible rage and mortification of her father, who had all the Scotch pride of family and pure blood.

This year we remained longer in the country than usual, and I went to spend Christmas with the Oswalds of Dunnikeir. The family consisted of a son, a colonel in the army, and three daughters, the youngest about my age, a bold horsewoman. She had talent, became a good Greek and Latin scholar, and was afterwards married to the Earl of Elgin. More than seventy years after this I had a visit from the Dean of Westminster and Lady Augusta Stanley, her daughter; a very charming person, who told me about her family, of which I had heard nothing for years. I was very happy to see the Dean, one of the most liberal and distinguished members of the Church of England, and son of my old friend the late Bishop of Norwich.

* * * * *

When I returned to Edinburgh Mr. Nasmyth was much pleased with the progress I had made in painting, for, besides having copied several landscapes he had lent me, I had taken the outline of a print and coloured it from a storm I saw at the end of our garden. This picture I still possess.

Dr. Blair, minister of the High Kirk of Edinburgh, the well-known author and professor of Rhetoric and Belles Lettres in the University, an intimate friend of my grandfather's, had heard of my turn for painting, and asked my mother to let him see some of my pictures. A few of the best were sent to him, and were returned after a few days accompanied by a long letter from the old gentleman, pointing out what he admired most in each picture. I was delighted with the letter, and not a little vain of the praise.

LETTER FROM DR. BLAIR TO MARY FAIRFAX.

My dear Miss Fairfax,

This comes to return you a thousand thanks for the pleasure and entertainment I have had from your landscape paintings. I had them placed in the best light I could contrive in my drawing-room, and entertained myself a good while every day looking at them and admiring their beauties, which always grew upon me. I intend to return them to you to-morrow, or rather on the beginning of next week; and as they were taken particular care of, I hope they shall not appear to have suffered any injury.

I have exhibited them to several people, some of whom were excellent judges, whom I brought on purpose to view them—Lady Miller, the Solicitor and Mrs. Blair, his lady, Dr. Hill, Miss Anne Ker of Nisbet, and a variety of ladies. All joined in praising them highly. The penserosa figure caught the highest admiration of any, from the gracefulness of the figure and attitude, and the boldness and propriety of the scenery. The two morning and evening views—one of Lochness, and the other of Elcho Castle—which make fine companions, and which I always placed together, were also highly admired. Each of them had their different partizans, and I myself was for a good while undetermined which of them to prefer. At last, I found the placidity of the scene in Elcho Castle, with the cottages among the trees, dwelt most on my imagination, though the gaiety and brightness of the morning sky in the other has also exquisite beauty. On the whole, I am persuaded that your taste and powers of execution in that art are uncommonly great, and that if you go on you must excel highly, and may go what length you please. Landscape painting has been always a great favourite with me; and you have really contributed much to my entertainment. As I thought you might wish to know my sentiments, after your paintings had been a little considered, I was led to write you these lines (in which I assure you there is nothing flattering), before sending back your pieces to you. With best compliments to Lady Fairfax, believe me,

Your obliged and most obedient Servant,

HUGH BLAIR.

ARGYLL SQUARE, 11*th April* (probably) 1796.

A day or two after this a Mrs. Ramsay, a rich proud widow, a relation of my mother's, came with her daughter, who was an heiress, to pay us a morning visit. Looking round the room she asked who had painted the pictures hung up on the walls. My mother, who was rather proud of them, said they were painted by me. "I am glad," said Mrs. Ramsay, "that Miss Fairfax has any kind of talent that may enable her to win her bread, for everyone knows she will not have a sixpence." It was a very severe hit, because it was true. Had it been my lot to win my bread by painting, I fear I should have fared badly, but I never should have been ashamed of it; on the contrary, I should have been very proud had I been successful. I must say the idea of making money had never entered my head in any of my pursuits, but I was intensely ambitious to excel in something, for I felt in my own breast that women were capable of taking a higher place in creation than that assigned to them in my early days, which was very low.

Not long after Mrs. Ramsay's visit to my mother, Miss Ramsay went to visit the Dons, at Newton Don, a pretty place near Kelso. Miss Ramsay and the three Miss Dons were returning from a long walk; they had reached the park of Newton Don, when they heard the dinner bell ring, and fearing to be too late for dinner, instead of going round, they attempted to cross a brook which runs through the park. One of the Miss Dons stumbled on the stepping-stones and fell into the water. Her two sisters and Miss Ramsay, trying to save her, fell in one after another. The three Miss Dons were drowned, but Miss Ramsay, who wore a stiff worsted petticoat, was buoyed up by it and carried down stream, where she caught by the branch of a tree and was saved. She never recovered the shock of the dreadful scene.

CHAPTER IV.

EDINBURGH SUPPER PARTIES—TOUR IN THE HIGHLANDS—MUTINY IN THE FLEET—BATTLE OF CAMPERDOWN.

[By this time my mother was grown up, and extremely pretty. All those who knew her speak of her rare and delicate beauty, both of face and figure. They called her the "Rose of Jedwood." She kept her beauty to the last day of her life, and was a beautiful old woman, as she had been a lovely young one. She used to say, laughing, that "it was very hard no one ever thought of painting her portrait so long as she was young and pretty." After she became celebrated, various likenesses were taken of her, by far the best of which are a beautiful bust, modelled at Rome in 1844 by Mr. Lawrence Macdonald, and a crayon drawing by Mr. James Swinton, done in London in 1848. My mother always looked considerably younger than her age; even at ninety, she looked younger than some who were her juniors by several years. This was owing, no doubt, principally to her being small and delicate in face and figure, but also, I think, to the extreme youthfulness and freshness of both her heart and mind, neither of which ever grew old. It certainly was not due to a youthful style of dress, for she had perfect taste in such matters, as well as in other things; and although no one spent less thought or money on it than she, my mother was at all times both neatly and becomingly dressed. She never was careless; and her room, her papers, and all that belonged to her were invariably in the most beautiful order. My mother's recollections of this period of her life are as follows:—

AT that time Edinburgh was really the capital of Scotland; most of the Scotch families of distinction spent the winter there, and we had numerous acquaintances who invited me to whatever gaiety was going on. As my mother refused to go into society when my father was at sea, I had to find a chaperon; but I never was at a loss, for we were somehow related to the Erskine family, and the Countess of Buchan, an amiable old lady, was always ready to take charge of me.

It was under Lady Buchan's care that I made my first appearance at a ball, and my first dancing partner was the late Earl of Minto, then Mr. Gilbert Elliot, with whom I was always on very friendly terms, as well as with his family. Many other ladies were willing to take charge of me, but a chaperon was only required for the theatre, and concerts, and for balls in the public assembly rooms; at private balls the lady of the house was thought sufficient. Still, although I was sure to know everybody in the room, or nearly so, I liked to have some one with whom to enter and to sit beside. Few ladies kept carriages, but went in sedan chairs, of which there were stands in the principal streets. Ladies were generally attended by a man-servant, but I went alone, as our household consisted of two maid-servants only. My mother knew, however, that the Highlanders who carried me could be trusted. I was fond of dancing, and never without partners, and often came home in bright daylight. The dances were reels, country dances, and sometimes Sir Roger de Coverley.

[At this period, although busily engaged in studying painting at Nasmyth's academy, practising the piano five hours a day, and pursuing her more serious studies zealously, my mother went a good deal into society, for Edinburgh was a gay, sociable place, and many people who recollect her at that time, and some who were her dancing partners, have told me she was much admired, and a great favourite. They said she had a graceful figure, below the middle size, a small head, well set on her shoulders, a beautiful complexion, bright, intelligent eyes, and a profusion of soft brown hair. Besides the various occupations I have mentioned, she made all her own dresses, even for balls. These, however, unlike the elaborate productions of our day, were simply of fine India muslin, with a little Flanders lace. She says of her life in Edinburgh:—

Girls had perfect liberty at that time in Edinburgh; we walked together in Princes Street, the fashionable promenade, and were joined by our dancing partners. We occasionally gave little supper parties, and presented these young men to our parents as they came in. At these meetings we played at games, danced reels, or had a little music—never cards. After supper there were toasts, sentiments, and songs. There were always one or two hot dishes, and a variety of sweet things and fruit. Though I was much more at ease in society, now, I was always terribly put out when asked for a toast or a sentiment. Like other girls, I did not dislike a little quiet flirtation; but I never could speak across a table, or take a leading part in conversation. This diffidence was probably owing to the secluded life I led in my early youth. At this time I gladly took part in any gaiety that was going on, and spent the day after a ball in idleness and gossiping with my friends; but these were rare occasions, for the balls were not numerous, and I never lost sight of the main object of my life, which was to prosecute my studies. So I painted at Nasmyth's, played the usual number of hours on the piano, worked and conversed with my mother in the evening; and as we kept early hours, I rose at day-break, and after dressing, I wrapped myself in a blanket from my bed on account of the excessive cold—having no fire at that hour—and read algebra or the classics till breakfast time. I had, and still have, determined perseverance, but I soon found that it was in vain to occupy my mind beyond a certain time. I grew tired and did more harm than good; so, if I met with a difficult point, for example, in algebra, instead of poring over it till I was bewildered, I left it, took my work or some amusing book, and resumed it when my mind was fresh. Poetry was my great resource on these occasions, but at a later period I read novels, the "Old English Baron," the "Mysteries of Udolpho," the "Romance of the Forest," &c. I was very fond of ghost and witch stories, both of which were believed in by most of the common people and many of the better educated. I heard an old naval officer say that he never opened his eyes after he was in bed. I asked him why? and he replied, "For fear I should see something!" Now I did not actually believe in either ghosts or witches, but yet, when alone in the dead of the night, I have been seized with a dread of, I know not what. Few people will now understand me if I say I was *eerie,* a Scotch expression for superstitious awe. I have been struck, on reading the life of the late Sir David Brewster, with the influence the superstitions of the age and country had on both learned and unlearned. Sir David was one of the greatest philosophers of the day. He was only a year younger than I; we were both born in Jedburgh, and both were influenced by the superstitions of our age and country in a similar manner, for he confessed that, although he did not believe in ghosts, he was *eerie* when sitting up to a late hour in a lone house that was haunted. This is a totally different thing from believing in spirit-rapping, which I scorn.

We returned as usual to Burntisland, in spring, and my father, who was at home, took my mother and me a tour in the Highlands. I was a great admirer of Ossian's poems, and viewed the grand and beautiful scenery with awe; and my father, who was of a romantic disposition, smiled at my enthusiastic admiration of the eagles as they soared above the mountains. These noble birds are nearly extirpated; and, indeed, the feathered tribes, which were more varied and numerous in Britain than in any part of Europe, will soon disappear. They will certainly be avenged by the insects.

On coming home from the journey I was quite broken-hearted to find my beautiful goldfinch, which used to draw its water so prettily with an ivory cup and little chain, dead in its cage. The odious wretches of servants, to whose care I trusted it, let it die of hunger. My heart is deeply pained as I write this, seventy years afterwards.

* * * * *

In Fifeshire, as elsewhere, political opinions separated friends and disturbed the peace of families; discussions on political questions were violent and dangerous on account of the hard-drinking then so prevalent. At this time the oppression and cruelty committed in Great Britain were almost beyond endurance. Men and women were executed for what at the present day would only have been held to deserve a few weeks' or months' imprisonment.* Every liberal opinion was crushed, men were entrapped into the army by promises which were never kept, and press-gangs tore merchant seamen from their families, and forced them to serve in the navy, where they were miserably provided for. The severity of discipline in both services amounted to torture. Such was the treatment of the brave men on whom the safety of the nation depended! They could bear it no longer; a mutiny broke out in the fleet which had been cruising off the Texel to watch the movements of a powerful Dutch squadron. The

* The late Justice Coltman told us, when he and Lady Coltman came to see my father and mother at Siena, that he recollected when he first went the circuit seeing more than twenty people hanged at once at York, chiefly for horse-stealing and such offences.—EDITOR.

men rose against their officers, took the command, and ship after ship returned to England, leaving only a frigate and the "Venerable," commanded by Admiral Duncan, with my father as his flag-captain. To deceive the Dutch, they continued to make signals, as if the rest of the fleet were in the offing, till they could return to England; when, without delay, Admiral Duncan and my father went alone on board each ship, ordered the men to arrest the ringleaders, which was done, and the fleet immediately returned to its station off the Texel. At last, on the morning of the 11th October, 1797, the Dutch fleet came out in great force, and formed in line of battle; that is, with their broadsides towards our ships. Then Admiral Duncan said to my father, "Fairfax, what shall we do?"—"Break their line, sir, and draw up on the other side, where they will not be so well prepared."—"Do it, then, Fairfax." So my father signalled accordingly. The circumstances of the battle, which was nobly fought on both sides, are historical. Nine ships of the line and two frigates were taken, and my father was sent home to announce the victory to the Admiralty. The rejoicing was excessive; every town and village was illuminated; and the Administration, relieved from the fear of a revolution, continued more confidently its oppressive measures.

When Admiral Duncan came to London, he was made a Baron, and afterwards Earl of Camperdown; and, by an unanimous vote of the House of Commons, he received a pension or a sum of money, I forget which; my father was knighted, and made Colonel of Marines. Earl Spencer was First Lord of the Admiralty at the time, and Lady Spencer said to my father, "You ask for the promotion of your officers, but you never have asked a reward for yourself." He replied, "I leave that to my country." But his country did nothing for him; and at his death my mother had nothing to live upon but the usual pension of an Admiral's widow, of seventy-five pounds a-year. Our friends, especially Robert Ferguson, junior, of Raith, made various attempts to obtain an addition to it; but it was too late: Camperdown was forgotten.

I remember one morning going to Lord Camperdown's house in Edinburgh with my mother, to see a very large painting, representing the quarter-deck of the "Venerable," Admiral Duncan, as large as life, standing upright, and the Dutch Admiral, De Winter, presenting his sword to my father. Another representation of the same scene may be seen among the numerous pictures of naval battles which decorate the walls of the great hall at Greenwich Hospital. Many years afterwards I was surprised to see an engraving of this very picture in the public library at Milan. I did not know that one existed.

At a great entertainment given to Lord Duncan by the East India Company, then in great power, the President asked my father, who sat at his left hand, if he had any relation in India? He replied, "My eldest son is in the Company's military service." "Then," said the President, "he shall be a Writer, the highest appointment in my power to bestow." I cannot tell how thankful we were; for, instead of a separation of almost a lifetime, it gave hopes that my brother might make a sufficient fortune in a few years to enable him to come home. There was a great review of the troops at Calcutta, under a burning sun; my brother returned to the barracks, sun-struck, where he found his appointment, and died that evening, at the age of twenty-one.

* * * * *

[My mother has often told us of her heart-broken parting with this brother on his going to India. It was then almost for a lifetime, and he was her favourite brother, and the companion of her childhood. He must have been wonderfully handsome, judging from a beautifully-painted miniature which we have of him.

Public events became more and more exciting every day, and difficulties occurred at home. There had been bad harvests, and there was a great scarcity of bread ; the people were much distressed, and the manufacturing towns in England were almost in a state of revolution ; but the fear of invasion kept them quiet. I gloried in the brilliant success of our arms by land and by sea ; and although I should have been glad if the people had resisted oppression at home, when we were threatened with invasion, I would have died to prevent a Frenchman from landing on our coast. No one can imagine the intense excitement which pervaded all ranks at that time. Every one was armed, and, notwithstanding the alarm, we could not but laugh at the awkward, and often ridiculous, figures of our old acquaintances, when at drill in uniform. At that time I went to visit my relations at Jedburgh. Soon after my arrival, we were awakened in the middle of the night by the Yeomanry entering the town at full gallop. The beacons were burning on the top of the Cheviots and other hills, as a signal that the French had landed. When day came, every preparation was made ; but it was a false alarm.

The rapid succession of victories by sea and land was intensely exciting. We always illuminated our house, and went to the rocky bank in our southern garden to see the illumination of Edinburgh, Leith, and the shipping in the Roads, which was inexpressibly beautiful, though there was no gas in those times. It often happened that balls were given by the officers of the ships of war that came occasionally to Leith Roads, and I was always invited, but never allowed to go; for my mother thought it foolish to run the risk of crossing the Firth, a distance of seven miles, at a late hour, in a small open boat and returning in the morning, as the weather was always uncertain, and the sea often rough from tide and wind. On one occasion, my father was at home, and, though it was blowing hard, I thought he would not object to accepting the invitation; but he said, "Were it a matter of duty, you should go, even at the risk of your life, but for a ball, certainly not."

We were as poor as ever, even more so; for my father was led into unavoidable expenses in London; so, after all the excitement, we returned to our more than usually economical life. No events worth mentioning happened for a long time. I continued my diversified pursuits as usual; had they been more concentrated, it would have been better; but there was no choice; for I had not the means of pursuing any one as far as I could wish, nor had I any friend to whom I could apply for direction or information. I was often deeply depressed at spending so much time to so little purpose.

CHAPTER V.

FIRST MARRIAGE (1804)—WIDOWHOOD—STUDIES—SECOND MARRIAGE.

[Mr. Samuel Greig was a distant relation of the Charters family. His father, an officer in the British navy, had been sent by our government, at the request of the Empress Catharine, to organize the Russian navy. Mr. Greig came to the Firth of Forth on board a Russian frigate, and was received by the Fairfaxes at Burntisland with Scotch hospitality, as a cousin. He eventually married my mother; not, however, until he had obtained the Russian consulship, and settled permanently in London, for Russia was then governed in the most arbitrary and tyrannical manner, and was neither a safe nor a desirable residence, and my grandfather only gave his consent to the marriage on this condition. My mother says:—

My cousin, Samuel Greig, commissioner of the Russian navy, and Russian consul for Britain, came to pay us a visit, and ultimately became my husband. Fortune I had none, and my mother could only afford to give me a very moderate trousseau, consisting chiefly of fine personal and household linen. When I was going away she gave me twenty pounds to buy a shawl or something warm for the following winter. I knew that the President of the Academy of Painting, Sir Arthur Shee, had painted a portrait of my father immediately after the battle of Camperdown, and I went to see it. The likeness pleased me,—the price was twenty pounds; so instead of a warm shawl I bought my father's picture, which I have since given to my nephew, Sir William George Fairfax. My husband's brother, Sir Alexis Greig, who commanded the Russian naval force in the Black Sea for more than twenty years, came to London about this time, and gave me some furs, which were very welcome. Long after this, I applied to Sir Alexis, at the request of Dr. Whewell, Master of Trinity College, Cambridge, and through his interest an order was issued by the Russian Government for simultaneous observations to be made of the tides on every sea-coast of the empire.

LETTER FROM DR. WHEWELL TO MRS. SOMERVILLE.

UNIVERSITY CLUB, *Jan.* 5, 1838.

MY DEAR MRS. SOMERVILLE,

I enclose a memorandum respecting tide observations, to which subject I am desirous of drawing the attention of the Russian Government. Nobody knows better than you do how much remains to be done respecting the tides, and what important results any advance in that subject would have. I hope, through your Russian friends, you may have the means of bringing this memorandum to the notice of the administration of their navy, so as to lead to some steps being taken, in the way of directing observations to be made. The Russian Government has shown so much zeal in promoting science, that I hope it will not be difficult to engage them in a kind of research so easy, so useful practically, and so interesting in its theoretical bearing.

Believe me, dear Mrs. Somerville,
Very faithfully yours,
W. WHEWELL.

* * * * *

My husband had taken me to his bachelor's house in London, which was exceedingly small and ill ventilated. I had a key of the neighbouring square, where I used to walk. I was alone the whole of the day, so I continued my mathematical and other pursuits, but under great disadvantages; for although my husband did not prevent me from studying, I met with no sympathy whatever from him, as he had a very low opinion of the capacity of my sex, and had neither knowledge of nor interest in science of any kind. I took lessons in French, and learnt to speak it so as to be understood. I had no carriage, so went to the nearest church; but, accustomed to our Scotch Kirk, I never could sympathise with the coldness and formality of the service of the Church of England. However, I thought it my duty to go to church and join where I could in prayer with the congregation.

There was no Italian Opera in Edinburgh; the first time I went to one was in London as chaperone to Countess Catharine Woronzow, afterwards Countess of Pembroke, who was godmother to my eldest son. I sometimes spent the evening with her, and occasionally dined at the embassy; but went nowhere else till we became acquainted with the family of Mr. Thomson Bonar, a rich Russian merchant, who lived in great luxury at a beautiful villa at Chiselhurst, in the neighbourhood of London, which has since become the refuge of the ex-Emperor Napoleon the Third and the Empress Eugénie. The family consisted of Mr. and Mrs. Bonar,—kind, excellent people,—with two sons and a daughter, all grown up. We were invited from time to time to spend ten days or a fortnight with them, which I enjoyed exceedingly. I had been at a riding school in Edinburgh, and rode tolerably, but had little practice, as we could not afford to keep horses. On our first visit, Mrs. Bonar asked me if I would ride with her, as there was a good lady's horse to spare, but I declined. Next day I said, "I should like to ride with you." "Why did you not go out with me yesterday?" she asked. "Because I had heard so much of English ladies' riding, that I thought you would clear all the hedges and ditches, and that I should be left behind lying on the ground." I spent many pleasant days with these dear good people; and no words can express the horror I felt when we heard that they had been barbarously murdered in their bedroom. The eldest son and daughter had been at a ball somewhere near, and on coming home they found that one of the men-servants had dashed out the brains of both their parents with a poker. The motive remains a mystery to this day, for it was not robbery.

* * * * *

[After three years of married life, my mother returned to her father's house in Burntisland, a widow, with two little boys. The youngest died in childhood. The eldest was Woronzow Greig, barrister-at-law, late Clerk of the Peace for Surrey. He died suddenly in 1865, to the unspeakable sorrow of his family, and the regret of all who knew him.

I was much out of health after my husband's death, and chiefly occupied with my children, especially with the one I was nursing; but as

I did not go into society, I rose early, and, having plenty of time, I resumed my mathematical studies. By this time I had studied plane and spherical trigonometry, conic sections, and Fergusson's "Astronomy." I think it was immediately after my return to Scotland that I attempted to read Newton's "Principia." I found it extremely difficult, and certainly did not understand it till I returned to it some time after, when I studied that wonderful work with great assiduity, and wrote numerous notes and observations on it. I obtained a loan of what I believe was called the Jesuit's edition, which helped me. At this period mathematical science was at a low ebb in Britain; reverence for Newton had prevented men from adopting the "Calculus," which had enabled foreign mathematicians to carry astronomical and mechanical science to the highest perfection. Professors Ivory and de Morgan had adopted the "Calculus"; but several years elapsed before Mr. Herschel and Mr. Babbage were joint-editors with Professor Peacock in publishing an abridged translation of La Croix's "Treatise on the Differential and Integral Calculus." I became acquainted with Mr. Wallace, who was, if I am not mistaken, mathematical teacher of the Military College at Marlow, and editor of a mathematical journal published there.

I had solved some of the problems contained in it and sent them to him, which led to a correspondence, as Mr. Wallace sent me his own solutions in return. Mine were sometimes right and sometimes wrong, and it occasionally happened that we solved the same problem by different methods. At last I succeeded in solving a prize problem! It was a diophantine problem, and I was awarded a silver medal cast on purpose with my name, which pleased me exceedingly.

Mr. Wallace was elected Professor of Mathematics in the University of Edinburgh, and was very kind to me. When I told him that I earnestly desired to go through a regular course of mathematical and astronomical science, even including the highest branches, he gave me a list of the requisite books, which were in French, and consisted of Francœur's pure "Mathematics," and his "Elements of Mechanics," La Croix's "Algebra," and his large work on the "Differential and Integral Calculus," together with his work on "Finite Differences and Series," Biot's "Analytical Geometry and Astronomy," Poisson's "Treatise on Mechanics," La Grange's "Theory of Analytical Functions," Euler's "Algebra," Euler's "Isoperimetrical Problems" (in Latin), Clairault's "Figure of the Earth," Monge's "Application of Analysis to Geometry," Callet's "Logarithms,"

La Place's "Mécanique Céleste," and his "Analytical Theory of Probabilities," &c., &c., &c.*

I was thirty-three years of age when I bought this excellent little library. I could hardly believe that I possessed such a treasure when I looked back on the day that I first saw the mysterious word "Algebra," and the long course of years in which I had persevered almost without hope. It taught me never to despair. I had now the means, and pursued my studies with increased assiduity; concealment was no longer possible, nor was it attempted. I was considered eccentric and foolish, and my conduct was highly disapproved of by many, especially by some members of my own family, as will be seen hereafter. They expected me to entertain and keep a gay house for them, and in that they were disappointed. As I was quite independent, I did not care for their criticism. A great part of the day I was occupied with my children; in the evening I worked, played piquet with my father, or played on the piano, sometimes with violin accompaniment.

* * * * *

This was the most brilliant period of the *Edinburgh Review*; it was planned and conducted with

* These books and all the other mathematical works belonging to my mother at the time of her death have been presented to the College for Women, at Girton, Cambridge.

consummate talent by a small society of men of the most liberal principles. Their powerful articles gave a severe and lasting blow to the oppressive and illiberal spirit which had hitherto prevailed. I became acquainted with some of these illustrious men, and with many of their immediate successors. I then met Henry Brougham, who had so remarkable an influence on my future life. His sister had been my early companion, and while visiting her I saw her mother—a fine, intelligent old lady, a niece of Robertson the historian. I had seen the Rev. Sydney Smith, that celebrated wit and able contributor to the *Review*, at Burntisland, where he and his wife came for sea-bathing. Long afterwards we lived on the most friendly terms till their deaths. Of that older group no one was more celebrated than Professor Playfair. He knew that I was reading the "Mécanique Céleste," and asked me how I got on? I told him that I was stopped short by a difficulty now and then, but I persevered till I got over it. He said, "You would do better to read on for a few pages and return to it again, it will then no longer seem so difficult." I invariably followed his advice and with much success.

Professor Playfair was a man of the most varied accomplishments and of the highest scientific distinction. He was an elderly man when I first became acquainted with him, by no means good-looking, but with a benevolent expression, somewhat concealed by the large spectacles he always wore. His manner was gravely cheerful; he was perfectly amiable, and was both respected and loved, but he could be a severe though just critic. He liked female society, and, philosopher as he was, marked attention from the sex obviously flattered him.

I had now read a good deal on the higher branches of mathematics and physical astronomy, but as I never had been taught, I was afraid that I might imagine that I understood the subjects when I really did not; so by Professor Wallace's advice I engaged his brother to read with me, and the book I chose to study with him was the "Mécanique Céleste." Mr. John Wallace was a good mathematician, but I soon found that I understood the subject as well as he did. I was glad, however, to have taken this resolution, as it gave me confidence in myself and consequently courage to persevere. We had advanced but little in this work when my marriage with my cousin, William Somerville (1812), put an end to scientific pursuits for a time.

CHAPTER VI.

SOMERVILLE FAMILY — DR. SOMERVILLE'S CHARACTER — LETTERS — JOURNEY TO THE LAKES—DEATH OF SIR WILLIAM FAIRFAX—REMINISCENCES OF SIR WALTER SCOTT.

[With regard to my father's family, I cannot do better than quote what my grandfather, the Rev. Thomas Somerville, says in his "Life and Times":—"I am a descendant of the ancient family of Somerville of Cambusnethan, which was a branch of the Somervilles of Drum, ennobled in the year 1424. Upon the death of George Somerville, of Corhouse, fifty years ago, I became the only male representative of the family." There is a quaint old chronicle, entitled "Memorie of the Somervilles," written by James, eleventh Lord Somerville, who died in 1690, which was printed for private distribution, and edited by Sir Walter Scott, and gives ample details of all the branches of our family. Although infinitely too prolix for our nineteenth century ideas, it contains many curious anecdotes and pictures of Scottish life.

My father was the eldest son of the minister of Jedburgh, and until his marriage with my mother, had lived almost entirely abroad and in our colonies. It was always a subject of regret to my mother that my father never could be induced to publish an account of his important travels in South Africa, for which he had ample materials in the notes he brought home, many of which we still possess. Without being very deeply learned on any one special subject, he was generally well-informed, and very intelligent. He was an excellent classical scholar, and could repeat long passages from Horace and other authors. He had a lively interest in all branches of natural history, was a good botanist and mineralogist, and could take note of all the strange animals, plants, or minerals he saw in his adventurous journies in the countries, now colonized, but then the hunting-grounds of Caffres and other uncivilized tribes. He was the first white man who penetrated so far into the country, and it was not without great risk. Indeed, on one occasion he was sentenced to death by a Caffre chief, and only saved by the interposition of the chief's mother.

My father's style in writing English was singularly pure and correct, and he was very fastidious on this topic—a severe critic, whether in correcting the children's lessons or in reading over the last proof sheets of my mother's works previous to their publication. These qualities would have fitted him very well to write the history of his travels, but he disliked the trouble of it, and, never having the slightest ambition on his own account, he let the time for publication slip by. Others travelled over the country he first explored, and the novelty was at an end. He was far happier in helping my mother in various ways, searching the libraries for the books she required, indefatigably copying and re-copying her manuscripts, to save her time. No trouble seemed too great which he bestowed upon her; it was a labour of love. My father was most kindhearted, and I have often heard my mother say how many persons he had assisted in life, and what generous actions he had done, many of them requited with ingratitude, and with betrayal of confidence. From the way my mother speaks of their life, it can be seen how happy was their marriage and how much sympathy there was between them. Speaking of his son's marriage with my mother, the Rev. Dr. Somerville says, in his "Life and Times," page 390: "To myself this connection was on every account peculiarly gratifying. Miss Fairfax had been born and nursed in my house; her father being at that time abroad on public service. She afterwards often resided in my family, was occasionally my scholar, and was looked upon by me and my wife as if she had been one of our own children. I can truly say, that next to them she was the object of our most tender regard. Her ardent thirst for knowledge, her assiduous application to study, and her eminent proficiency in science and the fine arts, have procured her a celebrity rarely obtained by any of her sex. But she never displays any pretensions to superiority, while the affability of her temper, and the gentleness of her manners afford constant sources of gratification to her friends. But what, above all other circumstances, rendered my son's choice acceptable to me, was that it had been the anxious, though secret, desire of my dear wife." I have already said that this esteem and affection of her father-in-law was warmly responded to by my mother. The following letter from her to him shows it vividly:—

LETTER FROM MRS. SOMERVILLE TO THE REV. DR. SOMERVILLE.

EDINBURGH, 1*st June*, 1812.

MY DEAR SIR,

I have this moment been gratified and delighted with your excellent and affectionate letter; the intercourse we have so long enjoyed has always been a source of the purest pleasure to me, and the kind interest you have taken from my infancy in my welfare was at all times highly flattering, and much valued; but now that the sacred name of Father is added, nothing is wanting to complete my happiness; and you may rest assured that William is not more anxious to hasten our visit to Jedburgh than I am. . . . . With the affectionate love of all here,

I remain your ever most affectionate daughter,

MARY SOMERVILLE.

P.S.—I am much flattered by the Latin quotation, and feel happy that your instructions have enabled me to read it.

* * * * *

[I will now proceed with the extracts from my mother's Recollections :—

My husband had been present at the taking of the Cape of Good Hope, and was sent by the authorities to make a treaty with the savage tribes on the borders of the colony, who had attacked the boors, or Dutch farmers, and carried off their cattle. In this journey he was furnished with a waggon and accompanied by Mr. Daniel, a good artist, who made drawings of the scenery, as well as of the animals and people. The savage tribes again became troublesome, and in a second expedition my cousin was only accompanied by a faithful Hottentot as interpreter. They were both mounted, and each led a spare horse with such things as were absolutely necessary, and when they bivouacked where, for fear of the natives, they did not dare light a fire to keep off the wild beasts, one kept watch while the other slept. After many adventures and dangers, my husband reached the Orange River, and was the first white man who had ever been in that part of Africa. He afterwards served in Canada and in Sicily at the head of the medical staff, under his friend General Sir James Craig. On returning to England he generally lived in London, so that he was seldom with his family, with whom he was not a favourite on account of his liberal principles, the very circumstance that was an attraction to me. He had lived in the world, was extremely handsome, had gentlemanly manners, spoke good English, and was emancipated from Scotch prejudices.

I had been living very quietly with my parents and children, so until I was engaged to my cousin I was not aware of the extreme severity with which my conduct was criticised by his family, and I have no doubt by many others; for as soon as our engagement was known I received a most impertinent letter from one of his sisters, who was unmarried, and younger than I, saying, she "hoped I would give up my foolish manner of life and studies, and make a respectable and useful wife to her brother." I was extremely indignant. My husband was still more so, and wrote a severe and angry letter to her; none of the family dared to interfere again. I lived in peace with her, but there was a coldness and reserve between us ever after. I forgot to mention that during my widowhood I had several offers of marriage. One of the persons whilst he was paying court to me, sent me a volume of sermons with the page ostentatiously turned down at a sermon on the Duties of a Wife, which were expatiated upon in the most illiberal and narrow-minded language. I thought this as impertinent as it was premature; sent back the book and refused the proposal.

My uncle, the Rev. Dr. Somerville, was delighted with my marriage with his son, for he was liberal, and sincerely attached to me. We were married by his intimate friend, Sir Henry Moncreiff Wellwood, and set off for the lakes in Cumberland. My husband's second sister, Janet, resolved to go with us, and she succeeded through the influence of my aunt, now my mother-in-law—a very agreeable, but bold, determined person, who was always very kind and sincerely attached to me. We were soon followed by my cousin, Samuel Somerville and his wife. We had only been a day or two in the little inn at Lowood when he was taken ill of a fever, which detained us there for more than a month. During his illness he took a longing for currant jelly, and here my cookery was needed; I made some that was excellent, and I never can forget the astonishment expressed at my being able to be so useful.

Somerville and I proceeded to London; and we managed to obtain a good position near Temple Bar to see the Emperor of Russia, the King of Prussia and his sons, Blucher, Platoff, the Hetman of the Cossacks, &c., &c., enter the City. There was a brilliant illumination in the evening, and great excitement. We often saw these noted persons afterwards, but we did not stay long in London, as my husband was appointed head of the Army Medical Department in Scotland, so we settled in Edinburgh. As he was allowed to have a secretary, he made choice of Donald Finlayson, a young man of great learning and merit, who was to act as tutor to my son, Woronzow Greig, then attending the High School, of which Mr. Pillans was master. Mr.

Finlayson was a remarkably good Greek scholar, and my husband said, "Why not take advantage of such an opportunity of improvement?" So I read Homer for an hour every morning before breakfast. Mr. Finlayson joined the army as surgeon, and distinguished himself by his courage and humanity during the battle of Waterloo; but he was lost in the march of the army to Paris, and his brother George, after having sought for him in vain, came to live with us in his stead. He excelled in botany, and here again, by my husband's advice, I devoted a morning hour to that science, though I was nursing a baby at the time. I knew the vulgar name of most of the plants that Mr. Finlayson had gathered, but now I was taught systematically, and afterwards made a herbarium, both of land plants and fuci. This young man's hopeful career was early arrested by his love of science, for he died of jungle fever in Bengal, caught while in search of plants.

Professor Playfair was now old, and resigned his chair, which Mr. Leslie was perfectly competent to fill on account of his acknowledged scientific acquirements; but, being suspected of heretical opinions, his appointment was keenly opposed, especially on the part of the clergy, and a violent contest arose, which ended in his favour. We became acquainted with him and liked him. He was a man of original genius, full of information on a variety of subjects, agreeable in conversation and good natured, but with a singular vanity as to personal appearance. Though one of the coarsest looking men I ever knew, he talked so much of polish and refinement that it tempted Mr. William Clerk, of Eldin, to make a very clever clay model of his ungainly figure. The professor's hair was grey, and he dyed it with something that made it purple; and, as at that time the art was not brought to its present perfection, the operation was tedious and only employed at intervals, so that the professor's hair was often white at the roots and dark purple at the extremities. He was always falling in love, and, to Somerville's inexpressible amusement, he made me his decoy duck, inviting me to see some experiments, which he performed dexterously; at the same time telling me to bring as many young ladies as I chose, especially Miss ——, for he was sure she had a turn for science. He was unfortunate in his aspirations, and remained a bachelor to the end of his life.

* * * * *

It was the custom in Edinburgh, especially among the clergy, to dine between the morning and evening service on Sundays, and to sup at nine or ten o'clock. In no family were these suppers more agreeable or cheerful than in that of Sir Henry Moncreiff

Wellwood, minister of the West Kirk. There were always a few of the friends of Sir Henry and Lady Moncreiff present, and we were invited occasionally. There was a substantial hot supper of roasted fowls, game, or lamb, and afterwards a lively, animated conversation on a variety of subjects, without a shade of austerity, though Sir Henry was esteemed an orthodox preacher.

There was an idiot in Edinburgh, the son of a respectable family, who had a remarkable memory. He never failed to go to the Kirk on Sunday, and on returning home could repeat the sermon word for word, saying, Here the minister coughed, Here he stopped to blow his nose. During the tour we made in the Highlands we met with another idiot who knew the Bible so perfectly that if you asked him where such a verse was to be found, he could tell without hesitation, and repeat the chapter. The common people in Scotland at that time had a kind of serious compassion for these harmless idiots, because "the hand of God was upon them."

The wise as well as the foolish are sometimes endowed with a powerful memory. Dr. Gregory, an eminent Edinburgh physician, one of the cleverest and most agreeable men I ever met with, was a remarkable instance of this. He wrote and spoke

Latin fluently, and Somerville, who was a good Latinist, met with a Latin quotation in some book he was reading, but not knowing from whence it was taken, asked his friend Dr. Gregory. "It is forty years since I read that author," said Dr. Gregory, "but I think you will find the passage in the middle of such a page." Somerville went for the book, and at the place mentioned there it was.

* * * * *

I had the grief to lose my dear father at this time. He had served sixty-seven years in the British Navy, and must have been twice on the North American station, for he was present at the taking of Quebec by General Wolfe, in 1759, and afterwards during the War of Independence. After the battle of Camperdown he was made a Colonel of Marines, and died, in 1813, Vice-Admiral of the Red.

* * * * *

Geology, which has now been so far advanced as a science, was still in its infancy. Professor Playfair and Mr. Hugh Miller had written on the subject; and in my gay young days, when Lady Helen Hall was occasionally my chaperone, I had heard that Sir James Hall had taken up the subject, but I did not care about it; I am certain that at that time I had never heard the word Geology. I think it was now, on going with Somerville to see the Edinburgh Museum, that I recognised the fossil plants I had seen in the coal limestone on the sands at the Links of Burntisland. Ultimately Geology became a favourite pursuit of ours, but then minerals were the objects of our joint study. Mineralogy had been much cultivated on the Continent by this time, especially in Germany. It had been established as a science by Werther, who was educated at an institution near the silver mines of Friburg, where he afterwards lectured on the properties of crystals, and had many pupils. In one of our tours on the Continent, Somerville and I went to see these silver mines and bought some specimens for our cabinet. The French took up the subject with great zeal, and the Abbé Haüy's work became a standard book on the science. Cabinets of minerals had been established in the principal cities of Great Britain, professors were appointed in the Universities, and collections of minerals were not uncommon in private houses. While quite a girl, I went with my parents to visit the Fergusons of Raith, near Kirkcaldy, and there I saw a magnificent collection of minerals, made by their son while abroad. It contained gems of great value and crystallized specimens of precious and other metals, which surprised and interested me, but seeing that such valuable things could never be obtained by me, I thought no more about them. In those early days I had every difficulty to contend with; now, through the kindness and liberal opinions of my husband, I had every encouragement. He took up the study of mineralogy with zeal, and I heartily joined with him. We made the acquaintance of Professor Jameson, a pupil of Werner's, whose work on mineralogy was of great use to us. We began to form a cabinet of minerals, which, although small, were good of their kind. We were criticized for extravagance, and, no doubt I had the lion's share of blame; but more of minerals hereafter.

* * * * *

Abbotsford is only twelve miles distant from Jedburgh, and my father-in-law, Dr. Somerville, and Sir Walter Scott had been intimate friends for many years, indeed through life. The house at Abbotsford was at first a mere cottage, on the banks of the Tweed; my brother-in-law, Samuel, had a villa adjacent to it, and John, Lord Somerville, had a house and property on the opposite bank of the river, to which he came every spring for salmon fishing. He was a handsome, agreeable man, had been educated in England, and as he thought he should never live in Scotland, he sold the family estate of Drum, within five miles of Edinburgh, which he afterwards regretted, and bought the property on the Tweed he then inhabited.

There was great intimacy between the three families, and the society was often enlivened by Adam Ferguson and Willie Clerk, whom we had met with at Raith. I shall never forget the charm of this little society, especially the supper-parties at Abbotsford, when Scott was in the highest glee, telling amusing tales, ancient legends, ghost and witch stories. Then Adam Ferguson would sing the "Laird of Cockpen," and other comic songs, and Willie Clerk amused us with his dry wit. When it was time to go away all rose, and, standing hand-in-hand round the table, Scott taking the lead, we sang in full chorus,

> Weel may we a' be,
> Ill may we never see ;
> Health to the king
> And the gude companie.

At that time no one knew who was the author of the Waverley Novels. There was much speculation and curiosity on the subject. While talking about one which had just been published, my son Woronzow said, "I knew all these stories long ago, for Mr. Scott writes on the dinner-table. When he has finished, he puts the green-cloth with the papers in a corner of the dining-room; and when he goes out, Charlie Scott and I read the stories." My son's tutor was the original of Dominie Sampson in "Guy Mannering." The "Memorie of the Somervilles" was edited by Walter Scott, from an ancient and very quaint manuscript found in the archives of the family, and from this he takes passages which he could not have found elsewhere. Although the work was printed it was never published, but copies were distributed to the different members of the family. One was of course given to my husband.

The Burning of the Water, so well described by Walter Scott in "Redgauntlet," we often witnessed. The illumination of the banks of the river, the activity of the men striking the salmon with the "leisters," and the shouting of the people when a fish was struck, was an animated, and picturesque, but cruel scene.

Sophia Scott, afterwards married to Mr. Lockhart, editor of the "Quarterly Review," was the only one of Sir Walter's family who had talent. She was not pretty, but remarkably engaging and agreeable, and possessed her father's joyous disposition as well as his memory and fondness for ancient Border legends and poetry. Like him, she was thoroughly alive to peculiarities of character, and laughed at them good-naturedly. She was not a musician, had little voice, but she sang Scotch songs and translations from the Gaelic with, or without, harp accompaniment; the serious songs with so much expression, and the merry ones with so much spirit, that she charmed everybody. The death of her brothers and of her father, to whom she was devotedly attached, cast a shade over the latter part of her life. Mr. Lockhart was clever and an able writer, but he was too sarcastic to be quite agreeable; however, we were always on the most friendly terms. He was of a Lanarkshire family and distantly related to Somerville. After the death of his wife and sons, Lockhart fell into bad health and lost much of his asperity.

Scott was ordered to go abroad for relaxation. Somerville and I happened to be at the seaport where he embarked, and we went to take leave of him. He kissed me, and said, "Farewell, my dear; I am going to die abroad like other British novelists." Happy would it have been if God had so willed it, for he returned completely broken down; his hopes were blighted, his sons dead, and his only remaining descendant was a grand-daughter, daughter of Mrs. Lockhart. She married Mr. James Hope, and soon died, leaving an only daughter, the last descendant of Sir Walter Scott. Thus the

"Merry, merry days that I have seen," ended very sadly.

* * * * *

When at Jedburgh, I never failed to visit James Veitch, who was Laird of Inchbonny, a small property beautifully situated in the valley of the Jed, at a short distance from the manse. He was a plough-wright, a hard-working man, but of rare genius, who taught himself mathematics and astronomy in the evenings with wonderful success, for he knew the motions of the planets, calculated eclipses and occultations, was versed in various scientific subjects, and made excellent telescopes, of which I bought a very small one; it was the only one I ever possessed. Veitch was handsome, with a singularly fine bald forehead and piercing eyes, that quite looked through one. He was perfectly aware of his talents, shrewd, and sarcastic. His fame had spread, and he had many visits, of which he was impatient, as it wasted his time. He complained especially of those from ladies not much skilled in science, saying, "What should they do but ask silly questions, when they spend their lives in doing naething but spatting muslin?" Veitch was strictly religious and conscientious, observing the Sabbath day with great solemnity; and I had the impression that he was stern to his wife, who seemed to be a person of intelligence, for I remember seeing her come from the washing-tub to point out the planet Venus while it was still daylight.

The return of Halley's comet, in 1835, exactly at the computed time, was a great astronomical event, as it was the first comet of long period clearly proved to belong to our system. I was asked by Mr. John Murray to write an article on the subject for the "Quarterly Review." After it was published, I received a letter from James Veitch, reproaching me for having mentioned that a peasant in Hungary was the first to see Halley's comet, and for having omitted to say that, "a peasant at Inchbonny was the first to see the comet of 1811, the greatest that had appeared for a century." I regretted, on receiving this letter, that I either had not known, or had forgotten the circumstance. Veitch has been long dead, but I avail myself of this opportunity of making the *amende honorable* to a man of great mental power and acquirements who had struggled through difficulties, unaided, as I have done myself.

LETTER FROM JAMES VEITCH TO MRS. SOMERVILLE.

INCHBONNY, 12*th October*, 1836.

DEAR MADAM,

I saw in the Quarterly review for December 1835 page 216 that the comet 1682 was discovered by a Peasent, George Palitzch residing in the neighbourhood of Dresden on the 25th of December 1758 with a small Telescope. But no mention is made of the Peasent at Inchbonny who first discovered the beautiful comet 1811. You will remember when Dr. Wollaston was at Inchbonny I put a difficult question to him that I could not solve about the focal distance of optic glasses when the Dr. got into a passion and said: Had he problems in his pocket ready to pull out in every occasion? and with an angry look at me said, You pretend to be the first that discovered the comet altho' it has been looked for by men of science for some time back. Now I never heard of such a thing and you will perhaps know something about it as the Dr. would not be mistaken. After we got acquainted, the Dr. was a warm friend of mine and I have often regretted that I had not improved the opportunity I had when he was here on many things he was master off. What ever others had known or expected I knew nothing about, But I know this, that on the 27th of August 1811 I first saw it in the NNW. part of the Heavens nigh the star marked 26 on the shoulder of the little Lion and continued treacing its path among the fixed stars untill it dissapeared and it was generally admitted that I had discovered it four days before any other person in Britain. However Mr. Thomas Dick on the Diffusion of Knowledge page

101 and 102 has made the following observation 'The splendid comet which appeared in our hemisphere in 1811 was first discovered in this country by a sawer. The name of this Gentleman is Mr. Veitch and I believe he resides in the neighbourhood of Kelso who with a Reflecting telescope of his own construction and from his sawpit as an observatory, descried that celestial visitant before it had been noticed by any other astronomer in North Britain.' A strange story—a sawer and a gentleman; and what is stranger still Mr. Baily would not have any place but the sawpit for his observatory on the 15th May last. I am sorry to say with all the improvement and learning that we can bost of in the present day Halley's comet the predictions have not been fulfilled, either with respect to time or place. Thus on the 10 October, at 50 minutes past 5 in the evening the Right ascension of the comet was 163° 37′, with 63° 38′ of north declination but by the nautical almanac for the 10 October its right ascension ought to have been 225° 2′ 6, and its declination 29° 33′. Hence the difference is no less than 61° in Right ascension and 34° in declination. When you have time, write me.

Dear Madam, I remain,

Yours sincerely,

JAMES VEITCH.

Sir David Brewster was many years younger than James Veitch; in his early years he assisted his father in teaching the parish-school at Jedburgh, and in the evenings he went to Inchbonny to study astronomy with James Veitch, who always called him Davie. They were as much puzzled about the meaning of the word parallax as I had been with regard to the word algebra, and only learnt what it meant when Brewster went to study for the kirk in Edinburgh. They were both very devout; nevertheless, Brewster soon gave up the kirk for science, and he devoted himself especially to optics, in which he made so many discoveries. Sir David was of ordinary height, with fair or sandy-coloured hair and blue eyes. He was by no means good-looking, yet with a very pleasant, amiable expression; in conversation he was cheerful and agreeable when quite at ease, but of a timid, nervous, and irritable temperament, often at war with his fellow-philosophers upon disputed subjects, and extremely jealous upon priority of discovery. I was much indebted to Sir David, for he reviewed my book on the "Connexion of the Physical Sciences," in the April number of the "Edinburgh Review" for 1834, and the "Physical Geography" in the April number of the "North British Review," both favourably.

CHAPTER VII

LIFE IN HANOVER SQUARE—VISIT TO FRANCE—ARAGO—CUVIER—ROME.

[My father was appointed, in 1816, a member of the Army Medical Board, and it became necessary for him to reside in London. He and my mother accordingly wished farewell to Scotland, and proceeded to take up their residence in Hanover Square. My mother preserved the following recollections of this journey :—

On our way we stopped a day at Birmingham, on purpose to see Watt and Boulton's manufactory of steam engines at Soho. Mr. Boulton showed us everything. The engines, some in action, although beautifully smooth, showed a power that was almost fearful. Since these early forms of the steam engine I have lived to see this all but omnipotent instrument change the locomotion of the whole civilized world by sea and by land.

Soon after our arrival in London we became acquainted with the illustrious family of the Herschels, through the kindness of our friend Professor Wallace, for it was by his arrangement that we spent a day with Sir William and Lady Herschel, at Slough. Nothing could exceed the kindness of Sir William. He made us examine his celebrated telescopes, and explained their mechanism ; and he showed us the manuscripts which recorded the numerous astronomical discoveries he had made. They were all arranged in the most perfect order, as was also his musical library, for that great genius was an excellent musician. Unfortunately, his sister, Miss Caroline Herschel, who shared in the talents of the family, was abroad, but his son, afterwards Sir John, my dear friend for many years, was at home, quite a youth. It would be difficult to name a branch of the physical sciences which he has not enriched by important discoveries. He has ever been a dear and valued friend to me, whose advice and criticism I gratefully acknowledge.

* * * * *

I took lessons twice a week from Mr. Glover, who painted landscapes very prettily, and I liked him on account of his kindness to animals, especially birds, which he tamed so that they flew before him when he walked, or else sat on the trees, and returned to him when he whistled. I regret now that I ever resumed my habit of painting in oil; water-colours are much better suited to an amateur, but as I had never seen any that were good, I was not aware of their beauty.

I also took lessons in mineralogy from Mrs. Lowry, a Jewess, the wife of an eminent line engraver, who had a large collection of minerals, and in the evening Somerville and I amused ourselves with our own, which were not numerous.

Our house in Hanover Square was within a walking distance of many of our friends, and of the Royal Institution in Albemarle Street, where I attended the lectures, and Somerville frequently went with me. The discoveries of Sir Humphry Davy made this a memorable epoch in the annals of chemical science. At this time there was much talk about the celebrated Count Rumford's steam kitchen, by which food was to be cooked at a very small expense of fuel. It was adopted by several people, and among others by Naldi, the opera singer, who invited some friends to dine the first day it was to be used. Before dinner they all went to see the new invention, but while Naldi was explaining its structure, it exploded and killed him on the spot. By this sad accident his daughter, a pretty girl and a good singer, was left destitute. A numerously-attended concert was given for her benefit, at which Somerville and I were present. She was soon after engaged to sing in Paris, but ultimately married the

Comte de Sparre, a French gentleman, and left the stage.

When MM. Arago and Biot came to England to continue the French arc of the meridian through Great Britain, they were warmly received by the scientific men in London, and we were always invited to meet them by those whom we knew. They had been told of my turn for science, and that I had read the works of La Place. Biot expressed his surprise at my youth.

*   *   *   *   *

One summer Somerville proposed to make a tour in Switzerland, so we set off, and on arriving at Chantilly we were told that we might see the château upon giving our cards to the doorkeeper. On reading our name, Mademoiselle de Rohan came to meet us, saying that she had been at school in England with a sister of Lord Somerville's, and was glad to see any of the family. She presented us to the Prince de Condé, a fine-looking old man, who received us very courteously, and sent the lord-in-waiting to show us the grounds, and especially the stables, the only part of the castle left in its regal magnificence after the Revolution. The Prince and the gentleman who accompanied us wore a gaudy uniform like a livery, which we were told was the Chantilly uniform, and that at each palace belonging to the Prince there was a different uniform worn by him and his court.

At Paris we were received with the kindest hospitality by M. and Mme. Arago. I liked her much, she was so gentle and ladylike; he was tall and good-looking, with an animated countenance and black eyes. His character was noble, generous, and singularly energetic; his manners lively and even gay. He was a man of very general information, and, from his excitable temperament, he entered as ardently into the politics and passing events of the time as into science, in which few had more extensive knowledge. On this account I thought his conversation more brilliant than that of any of the French savans with whom I was acquainted. They were living at the Observatory, and M. Arago showed me all the instruments of that magnificent establishment in the minutest detail, which was highly interesting at the time, and proved more useful to me than I was aware of. M. Arago made us acquainted with the Marquis de la Place, and the Marquise, who was quite an *élégante*. The Marquis was not tall, but thin, upright, and rather formal. He was distinguished in his manners, and I thought there was a little of the courtier in them, perhaps from having been so much at the court of the Emperor Napoleon, who had the highest regard for him.

Though incomparably superior to Arago in mathematics and astronomical science, he was inferior to him in general acquirements, so that his conversation was less varied and popular. We were invited to go early and spend a day with them at Arcœuil, where they had a country house. M. Arago had told M. de la Place that I had read the "Mécanique Céleste," so we had a great deal of conversation about astronomy and the calculus, and he gave me a copy of his "Système du Monde," with his inscription, which pleased me exceedingly. I spoke French very badly, but I was less at a loss on scientific subjects, because almost all my books on science were in French. The party at dinner consisted of MM. Biot, Arago, Bouvard, and Poisson. I sat next M. de la Place, who was exceedingly kind and attentive. In such an assemblage of philosophers I expected a very grave and learned conversation. But not at all! Everyone talked in a gay, animated, and loud key, especially M. Poisson, who had all the vivacity of a Frenchman. Madame Biot, from whom we received the greatest attention, made a party on purpose, as she said, to show us, "les personnes distinguées." Madame Biot was a well-educated woman, and had made a translation from the German of a work, which was published under the name of her husband. The dinner was very good, and Madame Biot was at great pains in placing every one. Those present were Monsieur and Madame Arago, Monsieur and Madame Poisson, who had only been married the day before, and Baron Humboldt. The conversation was lively and entertaining.

The consulate and empire of the first Napoleon was the most brilliant period of physical astronomy in France. La Grange, who proved the stability of the solar system, La Place, Biot, Arago, Bouvard, and afterwards Poinsot, formed a perfect constellation of undying names; yet the French had been for many years inferior to the English in practical astronomy. The observations made at Greenwich by Bradley, Maskelyn, and Pond, have been so admirably continued under the direction of the present astronomer-royal, Mr. Airy, the first practical astronomer in Europe, that they have furnished data for calculating the astronomical tables both in France and England.

The theatre was at this time very brilliant in Paris. We saw Talma, who was considered to be the first tragedian of the age in the character of Tancrède. I admired the skill with which he overcame the disagreeable effect which the rhyme of the French tragedies has always had on me. Notwithstanding his personal advantages, I thought him a great artist, though inferior to John Kemble. I am afraid my admiration of Shakespeare, my want of sympathy with the artificial style of French tragedy, and perhaps my youthful remembrance of our great tragedian Mrs. Siddons, made me unjust to Mademoiselle Duchênois, who, although ugly, was certainly an excellent actress and a favourite of the public. I was so fond of the theatre that I enjoyed comedy quite as much as tragedy, and was delighted with Mademoiselle Mars, whom we saw in Tartuffe. Some years later I saw her again, when, although an old woman, she still appeared handsome and young upon the stage, and was as graceful and lively as ever.

Soon after our dinner party at Arcœuil, we went to pay a morning visit to Madame de la Place. It was late in the day; but she received us in bed elegantly dressed. I think the curtains were of muslin with some gold ornaments, and the coverlet was of rich silk and gold. It was the first time that I had ever seen a lady receive in that manner. Madame de La Place was lively and agreeable; I liked her very much.

We spent a most entertaining day with M. and Madame Cuvier at the Jardin des Plantes, and saw the Museum, and everything in that celebrated establishment. On returning to the house, we found several people had come to spend the evening, and the conversation was carried on with a good deal of spirit; the Countess Albrizzi, a Venetian lady, of high acquirements, joined in it with considerable talent and animation. Cuvier had a very remarkable countenance, not handsome, but agreeable, and his manner was pleasing and modest, and his conversation very interesting. Madame de Staël having died lately, was much discussed. She was much praised for her good-nature, and for the brilliancy of her conversation. They agreed, that the energy of her character, not old age, had worn her out. Cuvier said, the force of her imagination misled her judgment, and made her see things in a light different from all the world. As a proof of this, he mentioned that she makes Corinne lean on a marble lion which is on a tomb in St. Peter's, at Rome, more than twenty feet high. Education was very much discussed. Cuvier said, that when he was sent to inspect the schools at Bordeaux and Marseilles, he found very few of the scholars who could perform a simple calculation in arithmetic; as to science, history, or literature, they were unknown, and the names of the most celebrated French philosophers, famed in other countries, were utterly unknown to those who lived in the provinces. M. Biot had written home, that he had found in Aberdeen not one alone, but many, who perfectly understood the object of his journey, and were competent to converse with him on the subject. Cuvier said such a circumstance constituted one of the striking differences between France and England; for in France science was highly cultivated, but confined to the capital. It was at M. Cuvier's that I first met Mr. Pentland, who made a series of physical and geological observations on the Andes of Peru. I was residing in Italy when I published my "Physical Geography," and Mr. Pentland* kindly undertook to carry the book through the press for me. From that time he has been a steady friend, ever ready to get me information, books, or anything I wanted. We became acquainted also with M. Gay-Lussac, who lived in the Jardin des Plantes, and with Baron Larrey, who had been at the head of the medical department of the army in Egypt under the first Napoleon.

* * * * *

At Paris I equipped myself in proper dresses, and we proceeded by Fontainebleau to Geneva, where we found Dr. Marcet, with whom my husband had

* Joseph Barclay Pentland, Consul-General in Bolivia (1836–39), died in London, July, 1873. He first discovered that Illimani and Sorata (not Chimborazo) were the highest mountains in America. (See Humboldt's "Kosmos.")

already been acquainted in London. I, for the first time, met Mrs. Marcet, with whom I have ever lived on terms of affectionate friendship. So many books have now been published for young people, that no one at this time can duly estimate the importance of Mrs. Marcet's scientific works. To them is partly owing that higher intellectual education now beginning to prevail among the better classes in Britain. They produced a great sensation, and went through many editions. Her "Conversations on Chemistry," first opened out to Faraday's mind that field of science in which he became so illustrious, and at the height of his fame he always mentioned Mrs. Marcet with deep reverence.

Through these kind friends we became acquainted with Professors De Candolle, Prevost, and De la Rive. Other distinguished men were also presented to us; among these was Mr. Sismondi, author of the "History of the Italian Republics." Madame Sismondi was a Miss Allen, of a family with whom we were very intimate.

[Some time after her return to England, my mother, desirous of continuing the study of botany, in which she had already attained considerable proficiency, wrote to M. De Candolle, asking his advice, and he sent her the following reply :—

LETTER FROM M. DE CANDOLLE TO MRS. SOMERVILLE.

LONDRES, 5 *Juin*, 1819.

MADAME,

Vous avez passé les premières difficultés de l'étude des plantes et vous me faites l'honneur de me consulter sur les moyens d'aller en avant ; connaissant votre goût et votre talent pour les sciences les plus relevées je ne craindrai point de vous engager à sortir de la Botanique élémentaire et à vous élever aux considérations et aux études qui en font une science susceptible d'idées générales, d'applications aux choses utiles et de liaison avec les autres branches des connaissances humaines. Pour cela il faut étudier non plus seulement la nomenclature et l'échafaudage artificiel qui la soutient, mais les rapports des plantes entre elles et avec les élémens extérieurs, ou en d'autres termes, la classification naturelle et la Physiologie.

Pour l'un et l'autre de ces branches de la science il est nécessaire en premier lieu de se familiariser avec la structure des plantes considérée dans leur caractère exacte. Vous trouverez un précis abrégé de ces caractères dans le 1er vol. de la Flore française ; vous la trouverez plus développé et accompagné de planches dans les Elémens de Botanique de Michel. Quant à la structure du fruit qui est un des points les plus difficiles et les plus importans, vous allez avoir un bon ouvrage traduit et augmenté par un de vos jeunes et habiles compatriotes, Mr. Lindley—c'est l'analyse du fruit de M. Richard. La traduction vaudra mieux que l'original. Outre ces lectures, ce qui vous apprendra surtout la structure des plantes, c'est de les analyser et de les décrire vous-même d'après les termes techniques; ce travail deviendrait pénible et inutile à faire sur un grand nombre de plantes, et il vaut mieux ne le faire que sur un très petit nombre d'espèces choisies dans des classes très distinctes. Quelques descriptions faites aussi complètes qu'il vous sera possible vous apprendra plus que tous les livres.

Dès que vous connaîtrez bien les organes et concurremment avec cette étude vous devrez chercher à prendre une idée de la classification naturelle. Je crains de vous paraître présomptueux en vous engageant à lire d'abord sous ce point de vue ma Théorie élémentaire. Après ces études ou à peu près en même temps pour profiter de la saison, vous ferez bien de rapporter aux ordres naturels toutes les plantes que vous aurez recueillies. La lecture des caractères des familles faites la plante à la main et l'acte de ranger vos plantes en familles vous feront connaître par théorie et par pratique ces groupes naturels. Je vous engage dans cette étude, surtout en le commencement, à ne donner que peu d'attention au système général qui lie les familles, mais beaucoup à la connaissance de la physionomie qui est propre à chacune d'elles. Sous ce point de vue vous pourrez trouver quelque intérêt à lire—1° les Tableaux de la Nature de M. de Humboldt; 2° mon essai sur les propriétés des plantes comparées avec leurs formes extérieures; 3° les remarques sur la géographie botanique de la Nouvelle Hollande et de l'Afrique, insérés par M. Robt. Brown à la fin du voyage de Finders et de l'expedition au Congo.

Quant à l'étude de la Physiologie ou de la connaissance des végétaux considérés comme êtres vivans, je vous engage à lire les ouvrages dans l'ordre suivant:

Philibert, Elémens de Bot. et de Phys., 3 vols.; la 2de partie des principes élémentaires de la Bot. de la Flore française. Vous trouverez la partie anatomique dans l'ouvrage de Mirbel; la partie chimique dans les recherches chimiques sur la Veget. de T. de Saussure; la partie statique dans la statique des végétaux de Hales, &c. &c. Mais je vous engage surtout à voir par vous-même les plantes à tous leurs ages, à suivre leur végétation, à les décrire en détail, en un mot à vivre avec elles plus qu'avec les livres.

Je désire, madame, que ces conseils puissent vous engager à suivre l'étude des plantes sous cette direction qui je crois en relève beaucoup l'importance et l'intérêt. Je m'estimerai heureux si en vous l'indiquant je puis concourir à vos succès futures et à vous initier dans une étude que j'ai toujours regardé comme une de celles qui peut le plus contribuer au bonheur journalier.

Je vous prie d'agréer mes hommages empressés.

De Candolle.

We had made the ordinary short tour through Switzerland, and had arrived at Lausanne on our way home, when I was taken ill with a severe fever which detained us there for many weeks. I shall never forget the kindness I received from two Miss Barclays, Quaker ladies, and a Miss Fotheringham, who, on hearing of my illness, came and sat up alternate nights with me, as if I had been their sister.

The winter was now fast approaching, and Somerville thought that in my weak state a warm climate was necessary; so we arranged with our friends, the Miss Barclays, to pass the Simplon together. We parted company at Milan, but we renewed our friendship in London.

We went to Monza, and saw the iron crown; and there I found the Magnolia grandiflora, which hitherto I had only known as a greenhouse plant, rising almost into a forest tree.

At Venice we renewed our acquaintance with the Countess Albrizzi, who received every evening. It was at these receptions that we saw Lord Byron, but he would not make the acquaintance of any English people at that time. When he came into the room I did not perceive his lameness, and thought him strikingly like my brother Henry, who was remarkably handsome. I said to Somerville, "Is Lord Byron like anyone you know?" "Your brother Henry, decidedly." Lord Broughton, then Sir John Cam Hobhouse, was also present.

At Florence, I was presented to the Countess of Albany, widow of Prince Charles Edward Stuart the Pretender. She was then supposed to be married to Alfieri the poet, and had a kind of state reception every evening. I did not like her, and never went again. Her manner was proud and insolent. "So you don't speak Italian; you must have had a very bad education, for Miss Clephane Maclane there [who was close by] speaks both French and Italian perfectly." So saying, she turned away, and never addressed another word to me. That evening I recognised in Countess Moretti my old friend Agnes Bonar. Moretti was of good family; but, having been banished from home for political opinions, he taught the guitar in London for bread, and an attachment was formed between him and his pupil. After the murder of her parents, they were both persecuted with the most unrelenting cruelty by her brother. They escaped to Milan where they were married.

I was still a young woman; but I thought myself too old to learn to speak a foreign language, consequently I did not try. I spoke French badly; and now, after several years' residence in Italy, although I can carry on a conversation fluently in Italian, I do not speak it well.

[When my mother first went abroad, she had no fluency in talking French, although she was well acquainted with the literature. To show how, at every period of her life, she missed no opportunity of acquiring information or improvement, I may mention that many years after, when we were spending a summer in Siena, where the language is spoken with great purity and elegance, she engaged a lady to converse in Italian with her for a couple of hours daily. By this means she very soon became perfectly familiar with the language, and could keep up conversation in Italian without difficulty. She never cared to write in any language but English. Her style has been reckoned particularly clear and good, and she was complimented on it by various competent judges, although she herself was always diffident about her writings, saying she was only a self-taught, uneducated Scotchwoman, and feared to use Scotch idioms inadvertently. In speaking she had a very decided but pleasant Scotch accent, and when aroused and excited, would often unconsciously use not only native idioms, but quaint old Scotch words. Her voice was soft and low, and her manner earnest.

On our way to Rome, where we spent the winter of 1817, it was startling to see the fine church of Santa Maria degli Angeli, below Assisi, cut in two; half of the church and half of the dome above it were still entire; the rest had been thrown down by the earthquake which had destroyed the neighbouring town of Foligno, and committed such ravages in this part of Umbria.

At that time I might have been pardoned if I had described St. Peter's, the Vatican, and the innumerable treasures of art and antiquity at Rome; but now that they are so well known it would be ridiculous and superfluous. Here I gained a little more knowledge about pictures; but I preferred sculpture, partly from the noble specimens of Greek art I saw in Paris and Rome, and partly because I was such an enthusiast about the language and everything belonging to ancient Greece. During this journey I was highly gratified, for we made the acquaintance of Thorwaldsen and Canova. Canova was gentle and amiable, with a beautiful countenance, and was an artist of great reputation. Thorwaldsen had a noble and striking appearance, and had more power and originality than Canova. His bas-reliefs were greatly admired. I saw the one he made of Night in the house of an English lady, who had a talent for modelling, and was said to be attached to him. We were presented to Pope Pius the Seventh; a handsome, gentlemanly, and amiable old man. He received us in a summer-house in the garden of the Vatican. He was sitting on a sofa, and made me sit beside him. His manners were simple and very gracious; he spoke freely of what he had suffered in France. He said, "God forbid that he should bear ill-will to any one; but the journey and the cold were trying to an old man, and he was glad to return to a warm climate and to his own country." When we took leave, he said to me, "Though a Protestant, you will be none the worse for an old man's blessing." Pius the Seventh was loved and respected; the people knelt to him as he passed. Many years afterwards we were presented to Gregory the Sixteenth, a very common-looking man, forming a great contrast to Pius the Seventh.

I heard more good music during this first visit to Rome than I ever did after; for besides that usual in St. Peter's, there was an Academia every week, where Marcello's Psalms were sung in concert by a number of male voices, besides other concerts, private and public. We did not make the acquaintance of any of the Roman families at this time; but we saw Pauline Borghese, sister of the Emperor Napoleon, so celebrated for her beauty, walking on the Pincio every afternoon. Our great geologist, Sir Roderick Murchison, with his wife, were among the English residents at Rome. At that time he hardly knew one stone from another. He had been an officer in the Dragoons, an excellent horseman, and a keen fox-hunter. Lady Murchison,—an amiable and accomplished woman, with solid acquirements which few ladies at that time possessed—had taken to the study of geology; and soon after her husband began that career which has rendered him the first geologist of our country. It was then that a friendship began between them and us, which will only end with life. Mrs. Fairfax, of Gilling Castle, and her two handsome daughters were also at Rome. She was my namesake—Mary Fairfax—and my valued friend till her death. Now, alas! many of these friends are gone.

There were such troops of brigands in the Papal States, that it was considered unsafe to go outside the gates of Rome. They carried off people to the mountains, and kept them till ransomed; sometimes even mutilated them, as they do at the present day in the kingdom of Naples. Lucien Bonaparte made a narrow escape from being carried off from his villa, Villa Ruffinella, near Frascati. When it could be proved that brigands had committed murder, they were confined in prisons in the Maremma, at Campo Morto, where fever prevails, and where they were supposed to die of malaria. I saw Gasperone, the chief of a famous band, in a prison at Civita Vecchia; he was said to be a relative of Cardinal Antonelli, both coming from the brigand village of Sonnino, in the Volscian mountains. In going to Naples our friends advised us to take a guard of soldiers; but these were suspected of being as bad, and in league with the brigands. So we travelled post without them; and though I foolishly insisted on going round by the ruins of ancient Capua, which was considered very unsafe, we arrived at Naples without any encounter. Here we met with the son and daughter of Mr. Smith, of Norwich, a celebrated leader in the anti-slavery question. This was a bond of interest between his family and me; for when I was a girl I took the anti-slavery cause so warmly to heart that I would not take sugar in my tea, or indeed taste anything with sugar in it. I was not singular in this, for my cousins and many of my acquaintances came to the same resolution. How long we kept it I do not remember. Patty Smith and I became great friends, and I knew her sisters; but only remember her niece Florence Nightingale as a very little child. My friend Patty was liberal in her opinions, witty, original, an excellent horsewoman, and drew cleverly; but from bad health she was peculiar in all her habits. She was a good judge of art. Her father had a valuable collection of pictures of the ancient masters; and I learnt much from her with regard to paintings and style in drawing. We went to see everything in Naples and its environs together, and she accompanied Somerville and me in an expedition to Pæstum, where we made sketches of the temples. At Naples we bought a beautiful cork model of the Temple of Neptune, which was placed on our mineral cabinet on our return to London. A lady who came to pay me a morning visit asked Somerville what it was; and when he told her, she said, "How dreadful it is to think that all the people who worshipped in that temple are in eternal misery, because they did not believe in our Saviour." Somerville asked, "How could they believe in Christ when He was not born till many centuries after?" I am sure she thought it was all the same.

* * * * *

There had been an eruption of Vesuvius just before our arrival at Naples, and it was still smoking very much; however, we ascended it, and walked round the crater, running and holding a handkerchief to our nose as we passed through the smoke, when the wind blew it to our side. The crater was just like an empty funnel, wide at the mouth, and narrowing to a throat. The lava was hard enough to bear us; but there were numerous *fumeroles*, or red-hot chasms, in it, which we could look into. Somerville bought a number of crystals from the guides, and went repeatedly to Portici afterwards to complete our collection of volcanic minerals.

They were excavating busily at Pompeii; at that time, and in one of our many excursions there Somerville bought from one of the workmen a bronze statuette of Minerva, and a very fine rosso antico Terminus, which we contrived to smuggle into Naples; and it now forms part of a small but excellent collection of antiques which I still possess. The excavations at that period were conducted with little regularity or direction, and the guides were able to carry on a contraband trade as mentioned. Since the annexation of the Neapolitan provinces to the kingdom of Italy, the Cavaliere Fiorelli has organized the system of excavations in the most masterly manner, and has made many interesting discoveries. About one-third of the town has been excavated since it was discovered till the present day.

In passing through Bologna, we became acquainted with the celebrated Mezzofanti, afterwards Cardinal. He was a quiet-looking priest; we could not see anything in his countenance that indicated talent, nor was his conversation remarkable; yet he told us that he understood fifty-two languages. He left no memoir at his death; nor did he ever trace any connection between these languages; it was merely an astonishing power, which led to nothing, like that of a young American I lately heard of, who could play eleven games at chess at the same time, without looking at any chess-board.

CHAPTER VIII.

EDUCATION OF DAUGHTERS—DR. WOLLASTON—DR. YOUNG—THE HERSCHELS.

When we returned to Hanover Square, I devoted my morning hours, as usual, to domestic affairs; but now my children occupied a good deal of my time. Although still very young, I thought it advisable for them to acquire foreign languages; so I engaged a French nursery-maid, that they might never suffer what I had done from ignorance of modern languages. I besides gave them instruction in such things as I was capable of teaching, and which were suited to their age.

It was a great amusement to Somerville and myself to arrange the minerals we had collected during our journey. Our cabinet was now very rich. Some of our specimens we had bought; our friends had given us duplicates of those they possessed; and George Finlayson, who was with our troops in Ceylon, and who had devoted all his spare time to the study of the natural productions of the country, sent us a valuable collection of crystals of sapphire, ruby, oriental topaz, amethyst, &c., &c. Somerville used to analyze minerals with the blowpipe, which I never did. One evening, when he was so occupied, I was playing the piano, when suddenly I fainted; he was very much startled, as neither I nor any of our family had ever done such a thing. When I recovered, I said it was the smell of garlic that had made me ill. The truth was, the mineral contained arsenic, and I was poisoned for the time by the fumes.

At this time we formed an acquaintance with Dr. Wollaston, which soon became a lasting friendship. He was gentlemanly, a cheerful companion, and a philosopher; he was also of agreeable appearance, having a remarkably fine, intellectual head. He was essentially a chemist, and discovered palladium; but there were few branches of science with which he was not more or less acquainted. He made experiments to discover imponderable matter; I believe, with regard to the ethereal medium. Mr. Brand, of the Royal Institution, enraged him by sending so strong a current of electricity through a machine he had made to prove electro-magnetic rotation, as to destroy it. His characteristic was extreme accuracy, which particularly fitted him for giving that precision to the science of crystallography which it had not hitherto attained. By the invention of the goniometer which bears his name, he was enabled to measure the angle formed by the faces of a crystal by means of the reflected images of bright objects seen in them. We bought a goniometer, and Dr. Wollaston, who often dined with us, taught Somerville and me how to use it, by measuring the angles of many of our crystals during the evening. I learnt a great deal on a variety of subjects besides crystallography from Dr. Wollaston, who, at his death, left me a collection of models of the forms of all the natural crystals then known.

Though still occasionally occupied with the mineral productions of the earth, I became far more interested in the formation of the earth itself. Geologists had excited public attention, and had shocked the clergy and the more scrupulous of the laity by proving beyond a doubt that the formation of the globe extended through enormous periods of time. The contest was even more keen then than it is at the present time about the various races of pre-historic men. It lasted very long, too; for after I had published my work on Physical Geography, I was preached against by name in York Cathedral. Our friend, Dr. Buckland, committed himself by taking the clerical view in his "Bridgewater Treatise;" but facts are such stubborn things, that he was obliged to join the geologists at last. He and Mrs. Buckland invited Somerville and me to spend a week with them in Christchurch College, Oxford. Mr. and Mrs. Murchison were their guests at the same time. Mr. Murchison (now Sir Roderick) was then rising rapidly to the pre-eminence he now holds as a geologist. We spent every day in seeing some of the numerous objects of interest in that celebrated university, venerable for its antiquity, historical records, and noble architecture.

Somerville and I used frequently to spend the evening with Captain and Mrs. Kater. Dr. Wollaston, Dr. Young, and others were generally of the party; sometimes we had music, for Captain and Mrs. Kater sang very prettily. All kinds of scientific subjects were discussed, experiments tried and astronomical observations made in a little garden in front of the house. One evening we had been trying the power of a telescope in separating double stars till about two in the morning; on our way home we saw a light in Dr. Young's window, and when Somerville rang the bell, down came the doctor himself in his dressing-gown, and said, "Come in; I have something curious to show you." Astronomical signs are frequently found on ancient Egyptian monuments, and were supposed to have been employed by the priests to record dates. Now Dr.

Young had received a papyrus from Egypt, sent to him by Mr. Salt, who had found it in a mummy-case; and that very evening he had proved it to be a horoscope of the age of the Ptolemies, and had determined the date from the configuration of the heavens at the time of its construction. Dr. Young had already made himself famous by the interpretation of hieroglyphic characters on a stone which had been brought to the British Museum from Rosetta in Egypt. On that stone there is an inscription in Hieroglyphics, the sacred symbolic language of the early Egyptians; another in the Enchorial or spoken language of that most ancient people, and a mutilated inscription in Greek. By the aid of some fragments of papyri Dr. Young discovered that the Enchorial language is alphabetical, and that nine of its letters correspond with ours; moreover, he discovered such a relation between the Enchorial and the hieroglyphic inscription that he interpreted the latter and published his discoveries in the years 1815 and 1816.

M. Champollion, who had been on the same pursuit, examined the fine collection of papyri in the museum at Turin, and afterwards went to Egypt to pursue his studies on hieroglyphics, to our knowledge of which he contributed greatly. It is to be regretted that one who had brought that branch of science to such perfection should have been so ungenerous as to ignore the assistance he had received from the researches of Dr. Young. When the Royal Institution was first established, Dr. Young lectured on natural philosophy. He proved the undulatory theory of light by direct experiment, but as it depended upon the hypothesis of an ethereal medium, it was not received in England, the more so as it was contrary to Newton's theory. The French *savans* afterwards did Young ample justice. The existence of the ethereal medium is now all but proved, since part of the corona surrounding the moon during a total solar eclipse is polarized—a phenomenon depending on matter. Young's Lectures, which had been published, were a mine of riches to me. He was of a Quaker family; but although he left the Society of Friends at an early age, he retained their formal precision of manner to the last. He was of a kindly disposition, and his wife and her sisters, with whom I was intimate, were much attached to him. Dr. Young was an elegant and critical scholar at a very early age; he was an astronomer, a mathematician, and there were few branches of science in which he was not versed. When young, his Quaker habits did not prevent him from taking lessons in music and dancing. I have heard him accompany his sister-in-law with the flute, while she played the piano. When not more than sixteen years of age he was so remarkable for steadiness and acquirements that he was engaged more as a companion than tutor to young Hudson Gurney, who was nearly of his own age. One spring morning Young came to breakfast in a bright green coat, and said in explanation of his somewhat eccentric costume for one who had been a Quaker, that it was suitable to the season. One day, on returning from their ride Gurney, leaped his horse over the stable-yard gate. Young, trying to do the same, was thrown; he got up, mounted, and made a second attempt with no better success; the third time he kept his seat, then quietly dismounting, he said, "What one man can do, another may."

* * * * *

One bright morning Dr. Wollaston came to pay us a visit in Hanover Square, saying, "I have discovered seven dark lines crossing the solar spectrum, which I wish to show you;" then, closing the window-shutters so as to leave only a narrow line of light, he put a small glass prism into my hand, telling me how to hold it. I saw them distinctly. I was among the first, if not the very first, to whom he showed these lines, which were the origin of the most wonderful series of cosmical discoveries, and have proved that many of the substances of our globe are also constituents of the sun, the stars, and even of the nebulæ. Dr. Wollaston gave me the little prism, which is doubly valuable, being of glass manufactured at Munich by Fraunhofer, whose table of dark lines has now become the standard of comparison in that marvellous science, the work of many illustrious men, brought to perfection by Bunsen and Kirchhoff.

* * * * *

Sir William Herschel had discovered that what appeared to be single stars were frequently two stars in such close approximation that it required a very high telescopic power to see them separately, and that in many of these one star was revolving in an orbit round the other. Sir James South established an observatory at Camden Hill, near Kensington, where he and Sir John Herschel united in observing the double stars and binary systems with the view of affording further data for improving our knowledge of their movements. In each two observations are requisite, namely, the distance between the two stars, and the angle of position, that is, the angle which the meridian or a parallel to the equator makes with the lines joining the two stars. These observations were made by adjusting a micrometer to a very powerful telescope, and were data sufficient for the determination of the orbit of the revolving star, should it be a binary system. I have given an account of this in the "Connexion of the Physical Sciences," so I shall only mention here that in one or two of the binary systems the revolving star has been seen to make more than one revolution, and that the periodical times and the elliptical elements of a great many other orbits have been calculated, though they are more than 200,000 times farther from the sun than we are.

After Sir John Herschel was married, we paid him a visit at Slough; fortunately, the sky was clear, and Sir John had the kindness to show me many nebulæ and clusters of stars which I had never seen to such advantage as in his 20 ft. telescope. I shall never forget the glorious appearance of Jupiter as he entered the field of that instrument.

For years the British nation was kept in a state of excitement by the Arctic voyages of our undaunted seamen in quest of a north-west passage from the Atlantic to the Pacific Ocean. The idea was not new, for a direct way to our Eastern possessions had been long desired. On this occasion the impulse was given by William Scoresby, captain of a whaler, who had sailed on the east coast of Greenland as high as the 80th parallel of latitude, and for two successive seasons had found that the sea between Greenland and Spitzbergen was free of ice for 18,000 square miles—a circumstance which had not occurred before in the memory of man. Scoresby was of rare genius, well versed in science, and of strict probity. When he published this discovery, the Admiralty, in the year 1818, sent off two expeditions, one under the command of Captains Franklin and Buchan to the east of Greenland, and another under Captains Ross and Parry to Baffin's Bay. Such was the beginning of a series of noble adventures, now the province of history.

I had an early passion for everything relating to the sea, and when my father was at home I never tired asking him questions about his voyages and the dangers to which he had been exposed. Now, when I knew something of nautical science, I entered with enthusiasm into the spirit of these Arctic voyages; nor was my husband less interested. We read Scoresby's whaling voyages with great delight, and we made the acquaintance of all the officers who had been on these northern expeditions.

Sir Edward Parry, who had brought us minerals and seeds of plants from Melville Island, invited us to see the ships prepared for his third voyage, and three years' residence in the Arctic seas. It is impossible to describe how perfectly everything was arranged: experience had taught them what was necessary for such an expedition. On this occasion

I put in practice my lessons in cookery by making a large quantity of orange marmalade for the voyage. When, after three years, the ships returned, we were informed that the name of Somerville had been given to an island so far to the north that it was all but perpetually covered with ice and snow. Notwithstanding the sameness which naturally prevails in the narratives of these voyages, they are invested with a romantic interest by the daring bravery displayed, and by the appalling difficulties overcome. The noble endeavour of Lady Franklin to save her gallant husband, and the solitary voyage of Sir Leopold McClintock in a small yacht in search of his lost friend, form the touching and sad termination to a very glorious period of maritime adventure. More than fifty years after these events I renewed my acquaintance with Lady Franklin. She and her niece came to see me at Spezia on their way to Dalmatia. She had circumnavigated the globe with her husband when he was governor in Australia. After his loss she and her niece had gone round the world a second time, and she assured me that although they went to Japan and China (less known at that time than they are now), they never experienced any difficulty. Seeing ladies travelling alone, people were always willing to help them. The French sent a Polar expedition under Captain Gaimard in the years 1838 and

1839 ; and the United States of North America took an active part in Arctic exploration. Whether Dr. Kane's discovery of an open polar ocean will ever be verified is problematical ; at all events, the deplorable fate of Sir John Franklin has put a stop to the chance of it for the present ; yet it is a great geographical question which we should all like to see decided.

Captain Sabine, of the Artillery (now General Sir Edward Sabine, President of the Royal Society), was appointed to accompany the first expedition under Captains Ross and Parry on account of his high scientific acquirements. The observations made during the series of Arctic voyages on the magnetism of the earth, combined with an enormous mass of observations made by numerous observers in all parts of the globe by sea and by land, have enabled Sir Edward Sabine, after a labour of nearly fifty years, to complete his marvellous system of terrestrial magnetism in both hemispheres. During that long period a friendship has lasted between Sir Edward and me. He has uniformly sent me copies of all his works ; to them I chiefly owe what I know on the subject, and quite recently I have received his latest and most important publication. Sir Edward married a lady of talent and scientific acquirements. She translated "Cosmos" from the German, and assisted and calculated for her husband in his laborious work.

I do not remember the exact period, but I think it was subsequent to the Arctic voyages, that the theory was discovered of those tropical hurricanes which cause such devastation by sea and land. Observations are now made on barometric pressure, and warnings are sent to our principal seaports by telegraph, as well as along both sides of the Channel; but notwithstanding numerous disastrous shipwrecks occur every winter on our dangerous coasts. They were far more numerous in my younger days. Life-boats were not then invented; now they are stationed on almost every coast of Great Britain, and on many continental shores. The readiness with which they are manned, and the formidable dangers encountered to save life, show the gallant, noble character of the sailor.

CHAPTER IX.

SOCIETY IN LONDON—CORONATION OF GEORGE IV.—LETTER TO DR. SOMERVILLE.

WE went frequently to see Mr. Babbage while he was making his Calculating-machines. He had a transcendant intellect, unconquerable perseverance, and extensive knowledge on many subjects, besides being a first-rate mathematician. I always found him most amiable and patient in explaining the structure and use of the engines. The first he made could only perform arithmetical operations. Not satisfied with that, Mr. Babbage constructed an analytical engine, which could be so arranged as to perform all kinds of mathematical calculations, and print each result.

Nothing has afforded me so convincing a proof of the unity of the Deity as these purely mental conceptions of numerical and mathematical science which have been by slow degrees vouchsafed to man, and are still granted in these latter times by the Differential Calculus, now superseded by the

Higher Algebra, all of which must have existed in that sublimely omniscient Mind from eternity.

Many of our friends had very decided and various religious opinions, but my husband and I never entered into controversy; we had too high a regard for liberty of conscience to interfere with any one's opinions, so we have lived on terms of sincere friendship and love with people who differed essentially from us in religious views, and in all the books which I have written I have confined myself strictly and entirely to scientific subjects, although my religious opinions are very decided.

Timidity of character, probably owing to early education, had a great influence on my daily life; for I did not assume my place in society in my younger days; and in argument I was instantly silenced, although I often knew, and could have proved, that I was in the right. The only thing in which I was determined and inflexible was in the prosecution of my studies. They were perpetually interrupted, but always resumed at the first opportunity. No analysis is so difficult as that of one's own mind, but I do not think I err much in saying that perseverance is a characteristic of mine.

* * * * *

Somerville and I were very happy when we lived in Hanover Square. We were always engaged in some pursuit, and had good society. General society was at that time brilliant for wit and talent. The Rev. Sidney Smith, Rogers, Thomas Moore, Campbell, the Hon. William Spencer, Macaulay, Sir James Mackintosh, Lord Melbourne, &c., &c., all made the dinner-parties very agreeable. The men sat longer at table than they do now, and, except in the families where I was intimate, the conversation of the ladies in the drawing-room, when we came up from dinner, often bored me. I disliked routs exceedingly, and should often have sent an excuse if I had known what to say. After my marriage I did not dance, for in Scotland it was thought highly indecorous for a married woman to dance. Waltzing, when first introduced, was looked upon with horror, and even in England it was then thought very improper.

One season I subscribed to the Concerts of Ancient Music, established by George the Third. They seemed to be the resort of the aged; a young face was scarcely to be seen. The music was perfect of its kind, but the whole affair was very dull. The Philharmonic Concerts were excellent for scientific musicians, and I sometimes went to them; but for my part I infinitely preferred hearing Pasta, Malibran, and Grisi, who have left the most vivid impression on my mind, although so different from each other. Somerville enjoyed a comic opera exceedingly, and so did I; and at that time Lablache was in the height of his fame. When Somerville and I made the tour in Italy already mentioned, we visited Catalani (then Madame Valabrèque) in a villa near Florence, to which she retired in her old age. She, however, died in Paris, of cholera, some years later.

Somerville liked the theatre as much as I did; so we saw all the greatest actors of the day, both in tragedy and comedy, and the English theatre was then excellent. Young, who was scarcely inferior to John Kemble, Macready, Kean, Liston, &c., and Miss O'Neill, who after a short brilliant career entered into domestic life on her marriage with Sir William Beecher, were all at the height of their fame. It was then I became acquainted with Lady Beecher, who was so simple and natural that no one could have discovered she had ever been on the stage. A very clever company of French comedians acted in a temporary theatre in Tottenham Court Road, where we frequently went with a party of friends, and enjoyed very pleasant evenings. I think my fondness for the theatre depended to a certain degree on my silent disposition; for unless among intimate friends, or when much excited, I was startled at the sound of my own voice in general conversation, from the shyness which has haunted me through life, and starts up occasionally like a ghost in my old age. At a play I was not called upon to make any exertion, but could enjoy at my ease an intellectual pleasure for the most part far superior to the general run of conversation.

Among many others, we were intimate with Dr. and Mrs. Baillie and his sisters. Joanna was my dear and valued friend to the end of her life. When her tragedy of "Montfort" was to be brought on the stage, Somerville and I, with a large party of her relations and friends, went with her to the theatre. The play was admirably acted, for Mrs. Siddons and her brother John Kemble performed the principal parts. It was warmly applauded by a full house, but it was never acted again. Some time afterwards "The Family Legend," founded on a Highland story, had better success in Edinburgh; but Miss Baillie's plays, though highly poetical, are not suited to the stage. Miss Mitford was more successful, for some of her plays were repeatedly acted. She excelled also as a writer. "Our Village" is perfect of its kind; nothing can be more animated than her description of a game of cricket. I met with Miss Austin's novels at this time, and thought them excellent, especially "Pride and Prejudice." It certainly formed a curious contrast to my old favourites, the Radcliffe novels and the ghost stories; but I had now come to years of discretion.

Among my Quaker friends I met with that amiable but eccentric person Mrs. Opie. Though a "wet" Quakeress, she continued to wear the peculiar dress. I was told that she was presented in it at the Tuileries, and astonished the French ladies. We were also acquainted with Mrs. Fry, a very different person, and heard her preach. Her voice was fine, her delivery admirable, and her prayer sublime. We were intimate with Mr. (now Sir Charles) Lyell, who, if I mistake not, first met with his wife at our house, where she was extremely admired as the beautiful Miss Horner. Until we lost all our fortune, and went to live at Chelsea, I used to have little evening parties in Hanover Square.

* * * * *

I was not present at the coronation of George the Fourth; but I had a ticket for the gallery in Westminster Hall, to see the banquet. Though I went very early in the morning, I found a wonderful confusion. I showed my ticket of admission to one official person after another; the answer always was "I know nothing about it." At last I got a good place near some ladies I knew; even at that early hour the gallery was full. Some time after the ceremony in the Abbey was over, the door of the magnificent hall was thrown open, and the king entered in the flowing curls and costume of Henry the Eighth, and, imitating the jaunty manner of that monarch, walked up the hall and sat down on the throne at its extremity. The peeresses had already taken their seats under the gallery, and the king was followed by the peers, and the knights of the Garter, Bath, Thistle, and St. Patrick, all in their robes. After every one had taken his seat, the Champion, on his horse, both in full armour, rode up the hall, and threw down a gauntlet before the king, while the heralds proclaimed that he was ready to do battle with any one who denied that George the Fourth was the liege lord of these realms. Then various persons presented offerings to the king in right of which they held their estates. One gentleman presented a beautiful pair of falcons in their hoods. While this pageantry and noise was at its height, Queen Caroline demanded to be admitted. There was a sudden silence and consternation,—it was like the "handwriting on the wall!" The sensation was intense. At last the order was given to refuse her admittance; the pageantry was renewed, and the banquet followed. The noise, heat, and vivid light of the illumination of the hall gave me a racking headache; at last I went out of the gallery and sat on a stair, where there was a little fresh air, and was very glad when all was over. Years afterwards I was present in Westminster Abbey at the coronation of our Queen, then a pretty young girl of eighteen. Placed in the most trying position at that early age, by her virtues, both public and private, she has endeared herself to the nation beyond what any sovereign ever did before.

* * * * *

I, who had so many occupations and duties at home, soon tired of the idleness and formality of visiting in the country. I made an exception, however, in favour of an occasional visit to Mr. Sotheby, the poet, and his family in Epping Forest, of which, if I mistake not, he was deputy-ranger; at all events, he had a pretty cottage there where he and his family received their friends with kind hospitality. He spent part of the day in his study, and afterwards I have seen him playing cricket with his son and grandson, with as much vivacity as any of them. The freshness of the air was quite reviving to Somerville and me; and our two little girls played in the forest all the day.

We also gladly went for several successive years to visit Sir John Saunders Sebright at Beechwood Park, Hertfordshire. Dr. Wollaston generally travelled with us on these occasions, when we had much conversation on a variety of subjects, scientific or general. He was remarkably acute in his observations on objects as we passed them. "Look at that ash tree; did you ever notice that the branches of the ash tree are curves of double curvature?" There was a comet visible at the time of one of these little journeys. Dr. Wollaston had made a drawing of the orbit and its elements; but, having left it in town, he described the lines so accurately without naming them, that I remarked at once, "That is the curtate or perihelion distance," which pleased him greatly, as it showed how accurate his description was. He was a chess-player, and, when travelling alone, he used to carry a book with diagrams of partially-played games, in which it is required to give checkmate in a fixed number of moves. He would study one of them, and then, shutting the book, play out the game mentally.

Although Sir John was a keen sportsman and fox-hunter in his youth, he was remarkable for his kindness to animals and for the facility with which he tamed them. He kept terriers, and his pointers were first rate, yet he never allowed his keepers to beat a dog, nor did he ever do it himself; he said a dog once cowed was good for nothing ever after. He trained them by tying a string to the collar and giving it a sharp pull when the dog did wrong, and patting him kindly when he did right. In this manner he taught some of his non-sporting dogs to play all sorts of tricks, such as picking out the card chosen by any spectator from a number placed in a circle on the floor, the signal being one momentary glance at the card, &c. &c. Sir John published a pamphlet on the subject, and sent copies of it to the sporting gentlemen and keepers in the county, I fear with little effect; men are so apt to vent their own bad temper on their dogs and horses.

At one of the battues at Beechwood, Chantrey killed two woodcocks at one shot. Mr. Hudson Gurney some time after saw a brace of woodcocks carved in marble in Chantrey's studio; Chantrey told him of his shot and the difficulty of finding a suitable inscription, and that it had been tried in Latin and even Greek without success. Mr. Gurney said it should be very simple, such as:—

Driven from the north, where winter starved them,
Chantrey first shot, and then he carved them.

Beechwood was one of the few places in Great Britain in which hawking was kept up. The falcons were brought from Flanders, for, except in the Isle of Skye, they have been extirpated in Great Britain like many other of our fine indigenous birds. Sir

John kept fancy pigeons of all breeds. He told me he could alter the colour of their plumage in three years by cross-breeding, but that it required fully six to alter the shape of the bird.

* * * * *

At some house where we were dining in London, I forget with whom, Ugo Foscolo, the poet, was one of the party. He was extremely excitable and irritable, and when some one spoke of a translation of Dante as being perfect, "Impossible," shouted Foscolo, starting up in great excitement, at the same time tossing his cup full of coffee into the air, cup and all, regardless of the china and the ladies' dresses. He died in England, I fear in great poverty. He was a most distinguished classical scholar as well as poet. His remains have been brought to Italy within these few years, and interred in Sante Croce, in Florence.

* * * * *

I had a severe attack of what appeared to be cholera, and during my recovery Mrs. Hankey very kindly lent us her villa at Hampstead for a few weeks. There I went with my children, Somerville with some friends always coming to dinner on the Sundays. On one of these occasions there was a violent thunderstorm, and a large tree was struck not far from the house. We all went to look at the tree as soon as the storm ceased, and found that a large mass of wood was scooped out of the trunk from top to bottom. I had occasion in two other instances to notice the same effect. Dr. Wollaston lent me a sextant and artificial horizon; so I amused myself taking the altitude of the sun, the consequence of which was that I became as brown as a mulatto, but I was too anxious to learn something of practical astronomy to care about the matter.

CHAPTER X.

DEATH OF MARGARET SOMERVILLE—LETTER FROM MRS. SOMERVILLE TO THE REV. DR. SOMERVILLE—LIFE AT CHELSEA—THE NAPIERS—MARIA EDGEWORTH—TOUR IN GERMANY.

OUR happy and cheerful life in Hanover Square came to a sad end. The illness and death of our eldest girl threw Somerville and me into the deepest affliction. She was a child of intelligence and acquirements far beyond her tender age.

[The long illness and death of this young girl fell very heavily on my mother, who by this time had lost several children. The following letter was written by her to my grandfather on this occasion. It shows her steadfast faith in the mercy and goodness of God, even when crushed by almost the severest affliction which can wring a mother's heart:—

MRS. SOMERVILLE TO THE REV. DR. SOMERVILLE.

LONDON, *October*, 1823.

MY DEAR FATHER,

I never was so long of writing to you, but when the heart is breaking it is impossible to find words adequate to its relief. We are in deep affliction, for though the first violence of grief has subsided, there has succeeded a calm sorrow not less painful, a feeling of hopelessness in this world which only finds comfort in the prospect of another, which longs for the consummation of all things that we may join those who have gone before. To return to the duties of life is irksome, even to those duties which were a delight when the candle of the Lord shone upon us. I do not arraign the decrees of Providence, but even in the bitterness of my soul I acknowledge the wisdom and goodness of God, and endeavour to be resigned to His will. It is ungrateful not to remember the many happy years we have enjoyed, but that very remembrance renders our present state more desolate and dreary—presenting a sad contrast. The great source of consolation is in the mercy of God and the virtues of those we lament; the full assurance that no good disposition can be lost but must be brought to perfection in a better world. Our business is to render ourselves fit for that blessed inheritance that we may again be united to those we mourn.

Your affectionate daughter,

MARY SOMERVILLE.

Somerville still held his place at the army medical board, and was now appointed physician to Chelsea Hospital; so we left our cheerful, comfortable house and went to reside in a government house in a very dreary and unhealthy situation, far from all our friends, which was a serious loss to me, as I was not a good walker, and during the whole time I lived at Chelsea I suffered from sick headaches. Still we were very glad of the appointment, for at this time we lost almost the whole of our fortune, through the dishonesty of a person in whom we had the greatest confidence.

All the time we lived at Chelsea we had constant intercourse with Lady Noel Byron and Ada, who lived at Esher, and when I came abroad I kept up a correspondence with both as long as they lived. Ada was much attached to me, and often came to stay with me. It was by my advice that she studied mathematics. She always wrote to me for an explanation when she met with any difficulty. Among my papers I lately found many of her notes, asking mathematical questions. Ada Byron married Lord King, afterwards created Earl of Lovelace, a college companion and friend of my son.

Somerville had formed a friendship with Sir Henry Bunbury when he had a command in Sicily, and we went occasionally to visit him at Barton in Suffolk. I liked Lady Bunbury very much; she was a niece of the celebrated Charles Fox, and had a turn for natural history. I had made a collection of native shells at Burntisland, but I only knew their vulgar names; now I learnt their scientific arrangement from Lady Bunbury. Her son, Sir Charles Bunbury, is an authority for fossil botany. The first Pinetum I ever saw was at Barton, and in 1837 I planted a cedar in remembrance of one of our visits.

Through Lady Bunbury we became intimate with all the members of the illustrious family of the Napiers, as she was sister of Colonel, afterwards General Sir William Napier, author of the "History of the Peninsular War." One day Colonel Napier, who was then living in Sloane Street, introduced Somerville and me to his mother, Lady Sarah Napier. Her manners were distinguished, and though totally blind, she still had the remains of great beauty; her hand and arm, which were exposed by the ancient costume she wore, were most beautiful still. The most sincere friendship existed between Richard Napier and his wife and me through life; I shall never forget their kindness to me at a time when I was in great sorrow. All the brothers are now gone. Richard and his wife were long in bad health, and he was nearly blind; but his wife never knew it, through the devoted attachment of Emily Shirriff, daughter of Admiral Shirriff, who was the comfort and consolation of both to their dying day.

Maria Edgeworth came frequently to see us when she was in England. She was one of my most intimate friends, warm-hearted and kind, a charming companion, with all the liveliness and originality of an Irishwoman. For seventeen years I was in constant correspondence with her. The cleverness and animation as well as affection of her letters I cannot express; certainly women are superior to men in letter-writing.

[The following is an extract from a letter from Maria Edgeworth to a friend concerning my mother :—

MARIA EDGEWORTH TO MISS . . . . .

BEECHWOOD PARK, *January* 17*th*, 1822.

We have spent two days pleasantly here with Dr. Wollaston, our own dear friend Mrs. Marcet, and the Somervilles. Mrs. Somerville is the lady who, Laplace says, is the only woman who understands his works. She draws beautifully, and while her head is among the stars her feet are firm upon the earth.

Mrs. Somerville is little, slightly made, fairish hair, pink colour, small, grey, round, intelligent, smiling eyes, very pleasing countenance, remarkably soft voice, strong, but well-bred Scotch accent; timid, not disqualifying timid, but naturally modest, yet with a degree of self-possession through it which prevents her being in the least awkward, and gives her all the advantage of her understanding, at the same time that it adds a prepossessing charm to her manner and takes off all dread of her superior scientific learning.

While in London I had a French maid for my daughters, and on coming to Chelsea I taught them a little geometry and algebra, as well as Latin and Greek, and, later, got a master for them, that they might have a more perfect knowledge of these languages than I possessed. Keenly alive to my own defects, I was anxious that my children should never undergo the embarrassment and mortification I had suffered from ignorance of the common European languages. I engaged a young German lady, daughter of Professor Becker, of Offenbach, near Frankfort, as governess, and was most happy in my choice; but after being with us for a couple of years, she had a very bad attack of fever, and was obliged to return home. She was replaced by a younger sister, who afterwards married Professor Trendelenburg, Professor of Philosophy at the University of Berlin. Though both these sisters were quite young, I had the most perfect confidence in them, from their strict conscientiousness and morality. They were well educated, ladylike, and so amiable, that they gained the friendship of my children and the affection of us all.

As we could with perfect confidence leave the children to Miss Becker's care, Sir James Mackintosh, Somerville and I made an excursion to the Continent. We went to Brussels, and what lady can go there without seeing the lace manufactory? I saw, admired,—and bought none! We were kindly received by Professor Quetelet, whom we had previously known, and who never failed to send me a copy of his valuable memoirs as soon as they were published. I have uniformly met with the greatest kindness from scientific men at home and abroad. If any of them are alive when this record is published, I beg they will accept of my gratitude. Of those that are no more I bear a grateful remembrance.

The weather was beautiful when we were at Brussels, and in the evening we went to the public garden. It was crowded with people, and very gay. We sat down, and amused ourselves by looking at them as they passed. Sir James was a most agreeable companion, intimate with all the political characters of the day, full of anecdote and historical knowledge. That evening his conversation was so brilliant that we forgot the time, and looking around found that everybody had left the garden, so we thought we might as well return to the hotel; but on coming to the iron-barred gate we found it locked. Sir James and Somerville begged some of those that were passing to call the keeper of the park to let us out; but they said it was impossible, that we must wait till morning. A crowd assembled laughing and mocking, till at last we got out through the house of one of the keepers of the park.

At Bonn we met with Baron Humboldt, and M. Schlegel, celebrated for his translation of Shakespeare. On going up the Rhine, Sir James knew the history of every place and of every battle that had been fought. A professor of his acquaintance in one of the towns invited us to dinner, and I was astonished to see the lady of the house going about with a great bunch of keys dangling at her side, assisting in serving up the dinner, and doing all the duty of carving, her husband taking no part whatever in it. I was annoyed that we had given so much trouble by accepting the invitation. In my younger days in Scotland, a lady might make the pastry and jelly, or direct in the kitchen; but she took no part in cooking or serving up the dinner, and never rose from the table till the ladies went to the drawing-room. However, as we could not afford to keep a regular cook, an ill-dressed dish would occasionally appear, and then my father would say, "God sends food, but the devil sends cooks."

In our tour through Holland, Somerville was quite at home, and amused himself talking to the people, for he had learnt the Dutch language at the Cape of Good Hope. We admired the pretty quaint costumes of the women; but I was the only one who took interest in the galleries. Many of the pictures of the Dutch school are very fine; but I never should have made a collection exclusively of them as was often done at one time in England. Lord Granville was British Minister at the Hague, and dining at the Embassy one day we met with a Mrs. ———, who, on hearing one of the attachés addressed as Mr. Abercromby,* said, "Pray, Lord Granville, is that a son of the great captain whom the Lord slew in the land of Egypt?"

I never met with Madame de Staël, but heard a great deal about her during this journey from Sir James Mackintosh, who was very intimate with her. At that time the men sat longer at table after dinner than they do now; and on one occasion, at a dinner party at Sir James's house, when Lady Mackintosh and the ladies returned to the drawing-room, Madame de Staël, who was exceedingly impatient of women's society, would not deign to enter into conversation with any of the ladies, but walked about the room; then suddenly ringing the bell, she said, "Ceci est insupportable!" and when the servant appeared, she said: "Tell your master to come upstairs directly; they have sat long enough at their wine."

* Afterwards Sir Ralph Abercromby, later Lord Dunfermline, minister first at Florence, then at Turin.

CHAPTER XI.

LETTER FROM LORD BROUGHAM — WRITES "MECHANISM OF THE HEAVENS"—ANECDOTE OF THE ROMAN IMPROVISATRICE—LETTERS FROM SIR JOHN HERSCHEL AND PROFESSOR WHEWELL—ELECTED HON. MEMBER OF THE ROYAL ASTRONOMICAL SOCIETY—NOTICE IN THE ACADÉMIE DES SCIENCES, AND LETTER FROM M. BIOT—PENSION—LETTER FROM SIR ROBERT PEEL—BEGINS TO WRITE ON THE CONNECTION OF THE PHYSICAL SCIENCES—VISIT TO CAMBRIDGE—LETTERS FROM PROFESSOR SEDGWICK AND LAPLACE.

[After my mother's return home my father received the following letter from Lord Brougham, which very importantly influenced the further course of my mother's life. It is dated March 27th, 1827 :—

LETTER FROM LORD BROUGHAM TO DR. SOMERVILLE.

MY DEAR SIR,

I fear you will think me very daring for the design I have formed against Mrs. Somerville, and still more for making you my advocate with her; through whom I have every hope of prevailing. There will be sent to you a prospectus, rules, and a preliminary treatise of our Society for Diffusing Useful Knowledge, and I assure you I speak without any flattery when I say that of the two subjects which I find it most difficult to see the chance of executing, there is one, which—unless Mrs. Somerville will undertake—none else can, and it must be left undone, though about the most interesting of the whole, I mean an account of the Mécanique Céleste ; the other is an account of the Principia, which I have some hopes of at Cambridge. The kind of thing wanted is such a description of that divine work as will both explain to the unlearned the sort of thing it is—the plan, the vast merit, the wonderful truths unfolded or methodized —and the calculus by which all this is accomplished, and will also give a somewhat deeper insight to the uninitiated. Two treatises would do this. No one without trying it can conceive how far we may carry ignorant readers into an understanding of the depths of science, and our treatises have about 100 to 800 pages of space each, so that one might give the more popular view, and another the analytical abstracts and illustrations. In England there are now not twenty people who know this great work, except by name ; and not a hundred who know it even by name. My firm belief is that Mrs. Somerville could add two cyphers to each of those figures. Will you be my counsel in this suit? Of course our names are concealed, and no one of our council but myself needs to know it.

Yours ever most truly,
H. Brougham.

My mother in alluding to the above says :—

This letter surprised me beyond expression. I thought Lord Brougham must have been mistaken with regard to my acquirements, and naturally concluded that my self-acquired knowledge was so far inferior to that of the men who had been educated in our universities that it would be the height of presumption to attempt to write on such a subject, or indeed on any other. A few days after this Lord Brougham came to Chelsea himself, and Somerville joined with him in urging me at least to make the attempt. I said, "Lord Brougham, you must be aware that the work in question never can be popularized, since the student must at least know something of the differential and integral calculi, and as a preliminary step I should have to prove various problems in physical mechanics and astronomy. Besides, La Place never gives diagrams or figures, because they are not necessary to persons versed in the calculus, but they would be indispensable in a work such as you wish me to write. I am afraid I am incapable of such a task: but as you both wish it so much, I shall do my very best upon condition of secrecy, and that if I fail the manuscript shall be put into the fire." Thus suddenly and unexpectedly the whole character and course of my future life was changed.

I rose early and made such arrangements with regard to my children and family affairs that I had time to write afterwards; not, however, without many interruptions. A man can always command his time under the plea of business, a woman is not allowed any such excuse. At Chelsea I was always supposed to be at home, and as my friends and acquaintances came so far out of their way on purpose to see me, it would have been unkind and ungenerous not to receive them. Nevertheless, I was sometimes annoyed when in the midst of a difficult problem some one would enter and say, "I have come to spend a few hours with you." However, I learnt by habit to leave a subject and resume it again at once, like putting a mark into a book I might be reading; this was the more necessary as there was no fire-place in my little room, and I had to write in the drawing-room in winter. Frequently I hid my papers as soon as the bell announced a visitor, lest anyone should discover my secret.

[My mother had a singular power of abstraction. When occupied with some difficult problem, or even a train of thought which deeply interested her, she lost all consciousness of what went on around her, and became so entirely absorbed that any amount of talking, or even practising scales and *solfeggi*, went on without in the least disturbing her. Sometimes a song or a strain of melody would recall her to a sense of the present, for she was passionately fond of music. A curious instance of this peculiarity of hers occurred at Rome, when a large party were assembled to listen to a celebrated improvisatrice. My mother was placed in the front row, close to the poetess, who, for several stanzas, adhered strictly to the subject which had been given to her. What it was I do not recollect, except that it had no connection with what followed. All at once, as if by a sudden inspiration, the lady turned her eyes full upon my mother, and with true Italian vehemence and in the full musical accents of Rome, poured forth stanza after stanza of the most eloquent panegyric upon her talents and virtues, extolling them and her to the skies. Throughout the whole of this scene, which lasted a considerable time, my mother remained calm and unmoved, never changing countenance, which surprised not only the persons present but ourselves, as we well knew how much she disliked any display or being brought forward in public. The truth was, that after listening for a while to the improvising, a thought struck her connected with some subject she was engaged in writing upon at the time and so entirely absorbed her that she heard not a word of all that had been declaimed in her praise, and was not a little surprised and confused when she was complimented on it. I call this, advisedly, a power of hers, for although it occasionally led her into strange positions, such as the one above mentioned, it rendered her entirely independent of outward circumstances, nor did she require to isolate herself from the family circle in order to pursue her studies. I have already mentioned that when we were very young she taught us herself for a few hours daily; when our lessons were over we always remained in the room with her, learning grammar, arithmetic, or some such plague of childhood. Any one who has plunged into the mazes of the higher branches of mathematics or other abstruse science, would probably feel no slight degree of irritation on being interrupted at a critical moment when the solution was almost within his grasp, by some childish question about tense or gender, or how much seven times seven made. My mother was never impatient, but explained our little difficulties quickly and kindly, and returned calmly to her own profound thoughts. Yet on occasion she could show both irritation and impatience—when we were stupid or inattentive, neither of which she could stand. With her clear mind she darted at the solution, sometimes forgetting that we had to toil after her laboriously step by step. I well remember her slender white hand pointing impatiently to the book or slate—"Don't you see it? there is no difficulty in it, it is quite clear." Things were so clear to her! I must here add some other recollections by my mother of this very interesting portion of her life.

I was a considerable time employed in writing this book, but I by no means gave up society, which would neither have suited Somerville nor me. We dined out, went to evening parties, and occasionally to the theatre. As soon as my work was finished I sent the manuscript to Lord Brougham, requesting that it might be thoroughly examined, criticised and destroyed according to promise if a failure. I was very nervous while it was under examination, and was equally surprised and gratified that Sir John Herschel, our greatest astronomer, and perfectly versed in the calculus, should have found so few errors. The letter he wrote on this occasion made me so happy and proud that I have preserved it.

LETTER FROM SIR JOHN HERSCHEL TO MRS. SOMERVILLE.

DEAR MRS. SOMERVILLE,

I have read your manuscript with the greatest pleasure, and will not hesitate to add, (because I am sure you will believe it sincere,) with the highest admiration. Go on thus, and you will leave a memorial of no common kind to posterity; and, what you will value far more than fame, you will have accomplished a most useful work. What a pity that La Place has not lived to see this illustration of his great work! You will only, I fear, give too strong a stimulus to the study of abstract science by this performance.

I have marked as somewhat obscure a part of the illustration of the principle of virtual velocities. . . . . . Will you look at this point again? I have made a trifling remark in page 6, but it is a mere matter of metaphysical nicety, and perhaps hardly worth pencilling your beautiful manuscript for.

Ever yours most truly,

J. HERSCHEL.

[In publishing the following letter, I do not consider that I am infringing on the rule I have followed in obedience to my mother's wishes, that is, to abstain from giving publicity to all letters which are of a private and confidential character. This one entirely concerns her scientific writings, and is interesting as showing the confidence which existed between Sir John Herschel and herself. This great philosopher was my mother's truest and best friend, one whose opinion she valued above all others, whose genius and consummate talents she admired, and whose beautiful character she loved with an intensity which is better shown by some extracts from her letters to be given presently than by anything I can say. This deep regard on her part he returned with the most chivalrous respect and admiration. In any doubt or difficulty it was his advice she sought, his criticism she submitted to; both were always frankly given without the slightest fear of giving offence, for Sir John Herschel well knew the spirit with which any remarks of his would be received.

FROM SIR JOHN HERSCHEL TO MRS. SOMERVILLE.

SLOUGH, *Feb. 23rd,* 1830.

MY DEAR MRS. SOMERVILLE,

. . . . . As you contemplate separate publication, and as the attention of many will be turned to a work from *your* pen who will just possess quantum enough of mathematical knowledge to be able to read the first chapter without being able to follow you into its application, and as these, moreover, are the very people who will think themselves privileged to criticise and use their privilege with the least discretion, I cannot recommend too much clearness, fulness, and order in the *exposé* of the principles. Were I you, I would devote to this first part at least double the space you have done. Your familiarity with the results and formulæ has led you into what is extremely natural in such a case—a somewhat hasty passing over what, to a beginner, would prove insuperable difficulties; and if I may so express it, a sketchiness of outline (as a painter you will understand my meaning, and what is of more consequence, see how it is to be remedied).

You have adopted, I see, the principle of virtual velocity, and the principle of d'Alembert, rather as separate and independent principles to be used as instruments of investigation than as convenient theories, flowing themselves from the general law of force and equilibrium, to be first *proved* and then remembered as compact statements in a form fit for use. The demonstration of the principle of virtual velocities is so easy and direct in Laplace that I cannot imagine anything capable of rendering it plainer than he has done. But a good deal more explanation of what *is* virtual velocity, &c., would be advantageous—and virtual velocities should be kept quite distinct from the arbitrary variations represented by the sign δ.

With regard to the *principle of d'Alembert*—take my advice and explode it altogether. It is the most awkward and involved statement of a plain dynamical equation that ever puzzled student. I speak feelingly and with a sense of irritation at the whirls and vortices it used to cause in my poor head when first I entered on this subject in my days of studentship. I know not a single case where its application does not create obscurity—nay *doubt*. Nor can a case ever occur where any such principle is called for. The general law that the change of motion is proportional to the moving force and takes place in its direction, provided we take care always to regard the *reaction* of curves, surfaces, obstacles, &c., as so many real moving forces of (for a time) unknown magnitude, will always help us out of any dynamical scrape we may get into. Laplace, page 20, Méc. Cél. art. 7, is a little obscure here, and in deriving his equation (*f*) a page of explanation would be well bestowed.

One thing let me recommend, if you use as principles either this, or that of virtual velocities, or any other, state them broadly and in general terms. . . . . You will think me, I fear, a rough critic, but I think of Horace's *good critic,*

> Fiet Aristarchus : nec dicet, cur ego amicum
> Offendam in nugis ? Hæ nugæ seria ducent
> In mala,

and what we can both now laugh at, and you may, if you like, burn as nonsense (I mean these remarks), would come with a very different kind of force from some sneering reviewer in the plenitude of his triumph at the detection of a slip of the pen or one of those little inaccuracies which *humana parum cavit natura.* . . . .

Very faithfully yours,

J. HERSCHEL.

[About the same time my father received a letter from Dr. Whewell, afterwards Master of Trinity College, Cambridge, dated 2nd November, 1831, in which he says :—

"I beg you to offer my best thanks to Mrs. Somerville for her kind present. I shall have peculiar satisfaction in possessing it as a gift of the author, a book which I look upon as one of the most remarkable which our age has produced, which would be highly valuable from anyone, and which derives a peculiar interest from its writer.

I am charged also to return the thanks of the Philosophical Society here for the copy presented to them. I have not thought it necessary to send the official letter containing the acknowledgment, as Mrs. Somerville will probably have a sufficient collection of specimens of such character. I have also to thank her on the part of our College for the copy sent to the library. I am glad that our young mathematicians in Trinity will have easy access to the book, which will be very good for them as soon as they can read it. When Mrs. Somerville shows herself in the field which we mathematicians have been labouring in all our lives, and puts us to shame, she ought not to be surprised if we move off to other ground, and betake ourselves to poetry. If the fashion of 'commendatory verses' were not gone by, I have no doubt her work might have appeared with a very pretty collection of well-deserved poetical praises in its introductory pages. As old customs linger longest in places like this, I hope she and you will not think it quite extravagant to send a single sonnet on the occasion.

"Believe me,

"Faithfully yours,

"W. Whewell."

TO MRS. SOMERVILLE,

ON HER "MECHANISM OF THE HEAVENS."

Lady, it was the wont in earlier days
When some fair volume from a valued pen,
Long looked for, came at last, that grateful men
Hailed its forthcoming in complacent lays:
As if the Muse would gladly haste to praise
That which her mother, Memory, long should keep
Among her treasures. Shall such usage sleep
With us, who feel too slight the common phrase

For our pleased thoughts of you, when thus we find
That dark to you seems bright, perplexed seems plain,
Seen in the depths of a pellucid mind,
Full of clear thought, pure from the ill and vain
That cloud the inward light? An honoured name
Be yours; and peace of heart grow with your growing fame.

[Professor Peacock, afterwards Dean of Ely, in a letter, dated February 14th, 1832, thanked my mother for a copy of the "Mechanism of the Heavens."

LETTER FROM PROFESSOR PEACOCK TO MRS. SOMERVILLE.

"I consider it to be a work which will contribute greatly to the extension of the knowledge of physical astronomy, in this country, and of the great analytical processes which have been employed in such investigations. It is with this view that I consider it to be a work of the greatest value and importance. Dr. Whewell and myself have already taken steps to introduce it into the course of our studies at Cambridge, and I have little doubt that it will immediately become an essential work to those of our students who aspire to the highest places in our examinations."

[On this my mother remarks:—

I consider this as the highest honour I ever received, at the time I was no less sensible of it, and was most grateful. I was surprised and pleased beyond measure to find that my book should be so much approved of by Dr. Whewell, one of the most eminent men of the age for science and literature; and by Professor Peacock, a profound mathematician, who with Herschel and Babbage had, a few years before, first introduced the calculus as an essential branch of science into the University of Cambridge.

In consequence of this decision the whole edition of the "Mechanism of the Heavens," amounting to 750 copies, was sold chiefly at Cambridge, with the exception of a very few which I gave to friends; but as the preface was the only part of the work that was intelligible to the general reader, I had some copies of it printed separately to give away.

I was astonished at the success of my book; all the reviews of it were highly favourable; I received letters of congratulation from many men of science. I was elected an honorary member of the Royal Astronomical Society at the same time as Miss Caroline Herschel. To be associated with so distinguished an astronomer was in itself an honour. Mr. De Morgan, to whom I am indebted for many excellent mathematical works, was then secretary of the society, and announced to us the distinction conferred. The council of the Society ordered that a copy of the "Greenwich Observations" should be regularly sent to me.

[The *Académie des Sciences* elected my mother's old friend M. Biot to draw up a report upon her "Mechanism of the Heavens," which he did in the most flattering terms, and upon my mother writing to thank him, replied as follows :—

FROM M. BIOT TO MRS. SOMERVILLE.

MADAME,

Revenu de Lyon depuis quelques jours, j'ai trouvé à Paris les deux lettres dont vous avez daigné m'honorer, et j'ai reçu également l'exemplaire de votre ouvrage que vous avez bien voulu joindre à la dernière. C'est être mille fois trop bonne, Madame, que de me remercier encore de ce qui m'a fait tant de plaisir. En rendant compte de cet étonnant Traité, je remplissais d'abord un devoir, puisque l'Académie m'avait chargé de le lire pour elle; mais ce devoir m'offrait un attrait que vous concevriez facilement, s'il vous était possible de vous rappeler l'admiration vive et profonde que m'inspira il y a longtems l'union si extraordinaire de tous les talens et de toutes les grâces, avec les connaissances sevères que nous autres hommes avions la folie de croire notre partage exclusif. Ce qui me charma alors, Madame, je n'ai pas cessé depuis de m'en souvenir; et des rapports d'amitié qui me sont bien chers, ont encore, à votre insçu, fortifié ces sentimens. Jugez donc, Madame, combien j'étais heureux d'avoir à peindre ce que je comprenais si bien, et ce que j'avais vu avec un si vif intérêt. Le plus amusant pour moi de cette rencontre, c'était de voir nos plus graves confrères, par exemple, Lacroix et Legendre, qui certes ne sont pas des esprits légers, ni galans d'habitude, ni faciles à émouvoir, me gourmander, comme ils le faisaient à chaque séance, de ce que je tardais tant à faire mon rapport, de ce que j'y mettais tant d'insouciance et si peu de grâce ; enfin, Madame, c'était une conquête intellectuelle complète. Je n'ai pas manqué de raconter cette circonstance comme un des fleurons de votre couronne. Je me suis ainsi acquitté envers eux ; et quant à vous, Madame, d'après la manière dont vous parlez vous-même de votre ouvrage, j'ai quelque espérance de l'avoir présenté sous le point de vue où vous semblez l'envisager. Mais, en vous rendant ce juste et sincère hommage et en l'insérant au Journal des Savans, je n'ai pas eu la précaution de demander qu'on m'en mit à part ; aujourd'hui que la collection est tirée je suis aux regrets d'avoir été si peu prévoyant. Au reste, Madame, il n'y a rien dans cet extrait que ce que pensent tous ceux qui vous connaissent, ou même qui ont eu une seule fois le bonheur de vous approcher. Vos amis trouveront que j'ai exprimé bien faiblement les charmes de votre esprit et de votre caractère ; charmes qu'ils doivent apprécier d'autant mieux qu'ils en jouissent plus souvent ; mais vous, Madame, qui êtes indulgente, vous pardonnerez la faiblesse d'un portrait qui n'a pu être fait que de souvenir.

J'ai l'honneur d'être, avec le plus profond respect,

Madame,

Votre très humble et très obéissant serviteur,

BIOT.

It was unanimously voted by the Royal Society of London, that my bust should be placed in their great Hall, and Chantrey was chosen as the sculptor. Soon after it was finished, Mr. Potter, a great shipbuilder at Liverpool, who had just completed a fine vessel intended for the China and India trade, wrote to my friend, Sir Francis Beaufort, hydrographer of the Royal Navy, asking him if I would give him permission to call her the "Mary Somerville," and to have a copy of my bust for her figure-head. I was much gratified with this, as might be expected. The "Mary Somerville" sailed, but was never heard of again; it was supposed she had foundered during a typhoon in the China sea.

I was elected an honorary member of the Royal Academy at Dublin, of the Bristol Philosophical Institution, and of the Société de Physique et d'Histoire Naturelle of Geneva, which was announced to me by a very gratifying letter from Professor Prevost.

Our relations and others who had so severely criticized and ridiculed me, astonished at my success, were now loud in my praise. The warmth with which Somerville entered into my success deeply affected me; for not one in ten thousand would have rejoiced at it as he did; but he was of a generous nature, far above jealousy, and he continued through life to take the kindest interest in all I did.

I now received the following letter from Sir Robert Peel, informing me in the handsomest manner that he had advised the King to grant me a pension of 200*l.* a year :—

LETTER FROM SIR ROBERT PEEL TO MRS. SOMERVILLE.

WHITEHALL GARDENS,
March, 1835.

MADAM,

In advising the Crown in respect to the grant of civil pensions, I have acted equally with a sense of public duty and on the impulse of my own private feelings in recognising among the first claims on the Royal favour those which are derived from eminence in science and literature.

In reviewing such claims, it is impossible that I can overlook those which you have established by the successful prosecution of studies of the highest order, both from the importance of the objects to which they relate, and from the faculties and acquirements which they demand.

As my object is a public one, to encourage others to follow the bright example which you have set, and to prove that great scientific attainments are recognised among public claims, I prefer making a direct communication to you, to any private inquiries into your pecuniary circumstances, or to any proposal through a third party. I am enabled to advise His Majesty to grant to you a pension on the civil list of two hundred pounds per annum; and if that provision will enable you to pursue your labours with less of anxiety, either as to the present or the future, I shall only be fulfilling a public duty, and not imposing upon you the slightest obligation, by availing myself of your permission to submit such a recommendation to the King.

I have the honour to be,
Madam, with the sincerest respect,
ROBERT PEEL.

I was highly pleased, but my pleasure was of short duration, for the very next day a letter informed us that by the treachery of persons in whom we trusted, the last remains of our capital were lost. By the kindness of Lord John Russell, when he was Prime Minister, a hundred a-year was added to my pension, for which I was very grateful.

* * * * *

After the "Mechanism of the Heavens" was published, I was thrown out of work, and now that I had got into the habit of writing I did not know what to make of my spare time. Fortunately the preface of my book furnished me with the means of active occupation; for in it I saw such mutual dependence and connection in many branches of science, that I thought the subject might be carried to a greater extent.

There were many subjects with which I was only partially acquainted, and others of which I had no previous knowledge, but which required to be carefully investigated, so I had to consult a variety of authors, British and foreign. Even the astronomical part was difficult, for I had to translate analytical formulæ into intelligible language, and to draw diagrams illustrative thereof, and this occupied the first seven sections of the book. I should have been saved much trouble had I seen a work on the subject by Mr. Airy, Astronomer-Royal, published subsequently to my book.

My son, Woronzow Greig, had been educated at Trinity College, Cambridge, and was travelling on the Continent, when Somerville and I received an invitation from the Principal, Dr. Whewell, to visit the University. Mr. Airy, then astronomer at Cambridge, now Astronomer-Royal at Greenwich, and Mrs. Airy kindly wished us to be their guests; but as the Observatory was at some distance from Cambridge, it was decided that we should have an apartment in Trinity College itself; an unusual favour where a lady is concerned. Mr. Sedgwick, the geologist, made the arrangements, received us, and we spent the first day at dinner with him. He is still alive*—one of my few coevals—either in Cambridge or England. The week we spent in Cambridge, receiving every honour from the heads of the University, was a period of which I have ever borne a proud and grateful remembrance.

* Professor Sedgwick died shortly after my mother.

[Professor Sedgwick wrote as follows to my father :—

FROM PROFESSOR SEDGWICK TO DR. SOMERVILLE.

TRINITY COLLEGE, *April*, 1834. 2.

MY DEAR SOMERVILLE,

Your letter delighted us. I have ordered dinner on Thursday at 6½ and shall have a small party to welcome you and Mrs. Somerville. In order that we may not have to fight for you, we have been entering on the best arrangements we can think of. On Tuesday you will, I hope, dine with Peacock; on Wednesday with Whewell; on Thursday at the Observatory. For Friday, Dr. Clarke, our Professor of Anatomy, puts in a claim. For the other days of your visit we shall, D.V., find ample employment. A four-poster bed now (a thing utterly out of our regular monastic system) will rear its head for you and Madame in the chambers immediately below my own; and your handmaid may safely rest her bones in a small inner chamber. Should Sheepshanks return, we can stuff him into a lumber room of the observatory; but of this there is no fear as I have written to him on the subject, and he has no immediate intention of returning. You will of course drive to the great gate of Trinity College, and my servant will be in waiting at the Porter's lodge to show you the way to your academic residence. We have no cannons at Trinity College, otherwise we would fire a salute on your entry; we will however give you the warmest greeting we can. Meanwhile give my best regards to Mrs. S.

And believe me most truly yours,

A. SEDGWICK.

* * * * *

La Place had a profound veneration for Newton; he sent me a copy of his "Système du Monde," and a letter, dated 15th August, 1824, in which he says: "Je publie successivement les divers livres du cinquième livre qui doit terminer mon traité de 'Mécanique Celeste,' et dans cela je donne l'analyse historique des recherches des géomètres sur cette matière, cela m'a fait relire avec une attention particulière l'ouvrage si incomparable des principes mathématiques de la philosophie naturelle de Newton, qui contient le germe de toutes ses recherches. Plus j'ai étudié cet ouvrage plus il m'a paru admirable, en me transportant surtout à l'époque où il a été publié. Mais en même tems que je sens l'élégance de la méthode synthétique suivant laquelle Newton a presenté ses découvertes, j'ai reconnu l'indispensable nécessité de l'analyse pour approfondir les questions très difficiles que Newton n'a pu qu'effleurer par la synthèse. Je vois avec un grand plaisir vos mathématiciens se livrer maintenant à l'analyse et je ne doute point qu'en suivant cette méthode avec la sagacité propre à votre nation ils ne seront conduits à d'importantes découvertes."

Newton himself was aware that by the law of gravitation the stability of the solar system was endangered. The power of analysis alone enabled La Grange to prove that all the disturbances arising from the reciprocal attraction of the planets and satellites are periodical, whatever the length of the periods may be, so that the stability of the solar system is insured for unlimited ages. The perturbations are only the oscillations of that immense pendulum of Eternity which beats centuries as ours beats seconds.

La Place, and all the great mathematicians of that period, had scarcely passed away when the more powerful Quaternion system began to dawn.

CHAPTER XII.

PARIS—ARAGO, LAFAYETTE, MM. BOUVARD, POISSON, LACROIX, &c., MARQUISE DE LA PLACE, DUPIN, F. COOPER—LEGITIMISTE SOCIETY —MAJENDIE—VISIT BARON LOUIS—LETTER FROM LAFAYETTE.

My health was never good at Chelsea, and as I had been working too hard, I became so ill, that change of air and scene were thought absolutely necessary for me. We went accordingly to Paris; partly, because it was near home, as Somerville could not remain long with us at a time, and, partly, because we thought it a good opportunity to give masters to the girls, which we could not afford to do in London. When we arrived, I was so weak, that I always remained in bed writing till one o'clock, and then, either went to sit in the Tuileries gardens, or else received visits. All my old friends came to see me, Arago, the first. He was more engaged in politics than science, and as party spirit ran very high at that time, he said he would send tickets of admission to the Chambers every time there was likely to be an "orage." When I told him what I was writing, he gave me some interesting memoirs, and lent me a mass of manuscripts, with leave to make extracts, which were very useful to me. General de La Fayette came to town on purpose to invite Somerville and me to visit him at La Grange, where we found him living like a patriarch, surrounded by his family to the fourth generation. He was mild, highly distinguished, and noble in his manners; his conversation was exceedingly interesting, as he readily spoke of the Revolution in which he had taken so active a part. Among other anecdotes, he mentioned, that he had sent the principal key of the Bastile to General Washington, who kept it under a glass case. He was much interested to hear that I could, in some degree, claim a kind of relationship with Washington, whose mother was a Fairfax. Baron Fairfax, the head of the family, being settled in America, had joined the independent party at the Revolution.

The two daughters of La Fayette, who had been in prison with him at Olmütz, were keen politicians, and discussed points with a warmth of gesticulation which amused Somerville and me, accustomed to our cold still manners. The grand-daughters, Mesdames de Rémusat and de Corcelles, were kind friends to me all the time I was in Paris.

M. Bouvard, whom we had known in London,

was now Astronomer-Royal of France, and he invited us to dine with him at the Observatory. The table was surrounded by savants, who complimented me on the "Mechanism of the Heavens." I sat next M. Poisson, who advised me in the strongest manner to write a second volume, so as to complete the account of La Place's works; and he afterwards told Somerville, that there were not twenty men in France who could read my book. M. Arago, who was of the party, said, he had not written to thank me for my book, because he had been reading it, and was busy preparing an account of it for the Journal of the Institute. At this party, I made the acquaintance of the celebrated astronomer, M. Pontécoulant, and soon after, of M. La Croix, to whose works I was indebted for my knowledge of the highest branches of mathematics. M. Prony, and M. Poinsot, came to visit me, the latter, an amiable and gentlemanly person; both gave me a copy of their works.

We had a long visit from M. Biot, who seemed really glad to renew our old friendship. He was making experiments on light, though much out of health; but when we dined with him and Madame Biot, he forgot for the time his bad health, and resumed his former gaiety. They made us promise to visit them at their country-house when we returned to England, as it lay on our road.

To my infinite regret, La Place had been dead some time; the Marquise was still at Arcœuil, and we went to see her. She received us with the greatest warmth, and devoted herself to us the whole time we were in Paris. As soon as she came to town, we went to make a morning visit; it was past five o'clock; we were shown into a beautiful drawing-room, and the man-servant, without knocking at the door, went into the room which was adjacent, and we heard her call out, "J'irai la voir! j'irai la voir!" and when the man-servant came out, he said, "Madame est désolée, mais elle est en chemise." Madame de La Place was exceedingly agreeable, the life of every party, with her cheerful gay manner. She was in great favour with the Royal Family, and was always welcome when she went to visit them in an evening. She received once a week, and her grand-daughter, only nineteen, lovely and graceful, was an ornament to her parties. She was already married to M. de Colbert, whose father fell at Corunna.

No one was more attentive to me than Dr. Milne-Edwards, the celebrated natural historian. He was the first Englishman who was elected a member of the Institute. I was indebted to him for the acquaintance of MM. Ampère and Becquerel. I believe Dr. Edwards was at that time writing on Physiology, and, in conversation, I happened to mention that the wild ducks in the fens, at Lincolnshire, always build their nests on high tufts of grass, or reeds, to save them from sudden floods; and that Sir John Sebright had raised wild ducks under a hen, which built their nests on tufts of grass as if they had been in the fens. Dr. Edwards begged of me to inquire for how many generations that instinct lasted.

Monsieur and Madame Gay Lussac lived in the Jardin des Plantes. Madame was only twenty-one, exceedingly pretty, and well-educated; she read English and German, painted prettily, and was a musician. She told me it had been computed, that if all the property in France were equally divided among the population, each person would have 150 francs a-year, or four sous per day; so that if anyone should spend eight sous a-day, some other person would starve.

The Duchesse de Broglie, Madame de Staël's daughter, called, and invited us to her receptions, which were the most brilliant in Paris. Every person of distinction was there, French or foreign, generally four or five men to one woman. The Duchess was a charming woman, both handsome and amiable, and received with much grace. The Duke was, then, Minister for Foreign Affairs. They were remarkable for their domestic virtues, as well as for high intellectual cultivation. The part the Duke took in politics is so well known, that I need not allude to it here.

At some of these parties I met with Madame Charles Dupin, whom I liked much. When I went to return her visit, she received us in her bed-room. She was a fashionable and rather elegant woman, with perfect manners. She invited us to dinner to meet her brother-in-law, the President of the Chamber of Deputies. He was animated and witty, very fat, and more ugly than his brother, but both were clever and agreeable. The President invited me to a very brilliant ball he gave, but as it was on a Sunday I could not accept the invitation. We went one evening with Madame Charles Dupin to be introduced to Madame de Rumford. Her first husband, Lavoisier, the chemist, had been guillotined at the Revolution, and she was now a widow, but had lived long separated from her second husband. She was enormously rich, and had a magnificent palace, garden, and conservatory, in which she gave balls and concerts. At all the evening parties in Paris the best bed-room was lighted up for reception like the other rooms. Madame de Rumford was capricious and ill-tempered ; however, she received me very well, and invited me to meet a very large party at dinner. Mr. Fenimore Cooper, the American novelist, with his wife and daughter, were among the guests. I found him extremely amiable and agreeable, which surprised me, for when I knew him in England he was so touchy that it was difficult to converse with him without giving him offence. He was introduced to Sir Walter Scott by Sir James Mackintosh, who said, in presenting him, "Mr. Cooper, allow me to introduce you to your great forefather in the art of fiction"; "Sir," said Cooper, with great asperity, "I have no forefather." Now, though his manners were rough, they were quite changed. We saw a great deal of him, and I was frequently in his house, and found him perfectly liberal; so much so, that he told us the faults of his country with the greatest frankness, yet he was the champion of America, and hated England.

None were kinder to us than Lord and Lady Granville. Lady Granville invited us to all her parties; and when Somerville was obliged to return to England, she assured him that in case of any disturbance, we should find a refuge in the Embassy. I went to some balls at the Tuileries with Madame de Lafayette Lasteyrie and her sister. The Queen Amélie was tall, thin, and very fair, not pretty, but infinitely more regal than Adelaide, Queen of

England, at that time. The Royal Family used to walk about in the streets of Paris without any attendants.

Sir Sydney Smith was still in Paris trying to renew the order of the Knights Templars. Somerville and I went with him one evening to a reception at the Duchesse d'Abrantés, widow of Junot. She was short, thick, and not in the least distinguished-looking, nor in any way remarkable. I had met her at the Duchesse de Broglie's, where she talked of Junot as if he had been in the next room. Sir Sydney was quite covered with stars and crosses, and I was amused with the way he threw his cloak back to display them as he handed me to the carriage.

I met with Prince Kosloffsky everywhere; he was the fattest man I ever saw, a perfect Falstaff. However, his intellect was not smothered, for he would sit an hour with me talking about mathematics, astronomy, philosophy, and what not. He was banished from Russia, and as he had been speaking imprudently about politics in Paris, he was ordered to go elsewhere; still, he lingered on, and was with me one morning when Pozzo di Borgo, the Russian Ambassador called. Pozzo di Borgo said to me, "Are you aware that Prince Kosloffsky has left Paris?" "Oh yes," I said, "I regret it much." He took the hint, and went away directly.

I had hitherto been entirely among the Liberal set. How it came that I was invited to dine with M. Héricourt de Thury, I do not remember. M. de Thury was simple in his manners, and full of information; he had been Director of the Mines under Napoleon, and had charge of the Public Buildings under Louis XVIII. and Charles X., but resigned his charges at the Revolution of July. At this time the Duchesse de Berry was confined in the citadel of Blaye. She had a strong party in Paris, who furiously resented the treatment she met with. M. de Thury was a moderate Legitimiste, but Madame was ultra. When I happened to mention that we had been staying with Lafayette, at La Grange, she was horrified, and begged of me not to talk politics, or mention where we had been, or else some of her guests would leave the room. The ladies of that party would not dance or go to any gay party; they had a part of the theatre reserved for themselves; they wore high dark dresses with long sleeves, called "Robes de Résistance," and even the Legitimiste newspapers appeared with black edges. They criticised those who gave balls, and Lady Granville herself did not escape their censure. The marriage of the Duchesse de Berry to the Marchese Lucchesi Palli made an immense sensation; it was discussed in the salons in a truly French manner; it was talked of in the streets; the Robes de Résistance were no longer worn, and the Legitimiste newspapers went out of mourning.

All parties criticised the British Administration in Ireland. A lady sitting by me at a party said, "No wonder so many English prefer France to so odious a country as England, where the people are oppressed, and even cabbages are raised in hot-beds." I laughed, and said, "I like England very well, for all that." An old gentleman, who was standing near us, said, "Whatever terms two countries may be on, it behoves us individuals to observe good manners;" and when I went away, this gentleman handed me to the carriage, though I had never seen him before.

The Marquise de La Place was commissioned by Dr. Majendie to invite me to meet her and Madame Gay Lussac at dinner. I was very unwilling to go; for I detested the man for his wanton cruelties, but I found I could not refuse on account of these ladies. There was a large party of *savants*, agreeable and gentlemanly; but Majendie himself had the coarsest manners; his conversation was horridly professional; many things were said and subjects discussed not fit for women to hear. What a contrast the refined and amiable Sir Charles Bell formed with Majendie! Majendie and the French school of anatomy made themselves odious by their cruelty, and failed to prove the true anatomy of the brain and nerves, while Sir Charles Bell did succeed, and thus made one of the greatest physiological discoveries of the age without torturing animals, which his gentle and kindly nature abhorred. To Lady Bell I am indebted for a copy of her husband's Life. She is one of my few dear and valued friends who are still alive.

* * * * *

While in Paris, I lost my dear mother. She died at the age of ninety, attended by my brother Henry. She was still a fine old lady, with few grey hairs. The fear of death was almost hereditary in the Charters family, and my mother possessed it in no small degree; yet when it came, she was perfectly composed and prepared for it. I have never had that fear; may God grant that I may be as calm and prepared as she was.

* * * * *

I was in better health, but still so delicate that I wrote in bed till one o'clock. The "Connexion of the Physical Sciences" was a tedious work, and the proof sheets had to be sent through the Embassy.

M. Arago told me that David, the sculptor,

wished to make a medallion of me; so he came and sat an hour with me, and pleased me by his intelligent conversation and his enthusiasm for art. A day was fixed, and he took my profile on slate with pink wax, in a wonderfully short time. He made me a present of a medallion in bronze, nicely framed, and two plaster casts for my daughters.

* * * * *

I frequently went to hear the debates in the Chambers, and occasionally took my girls, as I thought it was an excellent lesson in French. As party spirit ran very high, the scenes that occurred were very amusing. A member, in the course of his speech, happening to mention the word "liberté," the President Dupin rang the bell, called out "Stop, à propos de liberté," . . . jumped down from his seat, sprung into the tribune, pushed out the deputy, and made a long speech himself.

The weather being fine, we made excursions in the neighbourhood. At Sèvres I saw two pieces of china; on one of them was a gnu, on the other a zebra. Somerville had told me that soon after his return from his African expedition, he had given the original drawings to M. Brongniart then director of the manufactory.

Baron Louis invited me to spend a day with him and his niece, Mademoiselle de Rigny, at his country house, not far from Paris. I went with Madame de la Place, and we set out early, to be in time for breakfast. The road lay through the Forest of Vincennes. The Baron's park, which was close to the village of Petit-Brie, was very large, and richly wooded; there were gardens, hot-houses, and all the luxuries of an English nobleman's residence. The house was handsome, with a magnificent library; I remarked on the table the last numbers of the "Edinburgh" and "Quarterly" Reviews. Both the Baron and his niece were simple and kind. I was greatly taken with both; the Baron had all the quiet elegance of the old school, and his niece had great learning and the manners of a woman of fashion. She lived in perfect retirement, having suffered much in the time of the Revolution. They had both eventful lives; for Baron Louis, who had been in orders, and Talleyrand officiated at the Champs de Mars when Louis the Sixteenth took the oath to maintain the constitution. Field-Marshal Macdonald, Duc de Tarante, and his son-in-law, the Duc de Massa; Admiral de Rigny, Minister of Marine; M. Barthe, Garde des Sceaux; and the Bouvards, father and son, formed the party. After spending a most delightful and interesting day, we drove to Paris in bright moonlight.

Our friends in Paris and at La Grange had been

so kind to us that we were very sad when we went to express our gratitude and take leave of them. We only stayed two days at La Grange, and when we returned to Paris, Somerville went home and my son joined us, when we made a rapid tour in Switzerland, the only remarkable event of which was a singular atmospheric phenomenon we saw on the top of the Grimsel. On the clouds of vapour below us we saw our shadows projected, of giant proportions, and each person saw his own shadow surrounded by a bright circle of prismatic colours. It is not uncommon in mountain regions.

* * * * *

[General Lafayette and all his family were extremely kind to my mother. He was her constant visitor, and we twice visited him at his country house, La Grange. He wished to persuade my mother to go there for some days, after our return from Switzerland, which we did not accomplish. The General wrote the following letter to my father:—

FROM LAFAYETTE TO DR. SOMERVILLE.

La Grange, 31st *October*, 1833.

My dear Sir,

I waited to answer your kind letter, for the arrival of Mr. Coke's* precious gift, which nobody could higher value, on every account, than the grateful farmer on whom it has been bestowed. The heifers and bull

* Mr. Coke, of Holkham, afterwards Earl of Leicester.

are beautiful; they have reached La Grange in the best order, and shall be tenderly attended to. . . . It has been a great disappointment not to see Mrs. Somerville and the young ladies before their departure. Had we not depended on their kind visit, we should have gone to take leave of them. They have had the goodness to regret the impossibility to come before their departure. Be so kind as to receive the affectionate friendship and good wishes of a family who are happy in the ties of mutual attachment that bind us to you and them. . . . Public interest is now fixed upon the Peninsula, and while dynasties are at civil war, and despotic or *juste milieu* cabinets seem to agree in the fear of a genuine development of popular institutions, the matter for the friends of freedom is to know how far the great cause of Europe shall be forwarded by these royal squabbles.

We shall remain at La Grange until the opening of the session, hoping that, notwithstanding your and the ladies' absence, your attention will not be quite withdrawn from our interior affairs—the sympathy shall be reciprocal.

With all my heart, I am

Your affectionate friend,

LAFAYETTE.

CHAPTER XIII.

RETURN TO ENGLAND—LETTER FROM HALLAM—TREATISE ON THE FORM AND ROTATION OF THE EARTH AND PLANETS—SECOND EDITION OF "CONNEXION OF PHYSICAL SCIENCES"—LETTERS FROM MARIA EDGEWORTH, MISS BERRY, LORD BROUGHAM, MRS. MARCET, ADMIRAL SMYTH—DOUBLE STARS—ECLIPSE OF DOUBLE STARS—LETTER FROM ADMIRAL SMYTH—SIR WILLIAM HERSCHEL—NEBULÆ—LETTER FROM LORD ROSSE—LETTER FROM SIR JOHN HERSCHEL—SIR JAMES SOUTH'S OBSERVATORY—MR. JOHN MURRAY—MISS BERRY—LORD DUDLEY—MR. BOWDITCH AND OTHER DISTINGUISHED AMERICANS—MRS. BROWNING WASHINGTON—LETTER FROM THE REV. DR. TUCKERMAN—SIR WILLIAM FAIRFAX ATTACKED BY HIGHWAYMEN.

As soon as we returned to Chelsea, the "Connexion of the Physical Sciences" was published. It was dedicated to Queen Adelaide, who thanked me for it at a drawing-room. Some time after Somerville and I went to Scotland; we had travelled all night in the mail coach, and when it became light, a gentleman who was in the carriage said to Somerville, "Is not the lady opposite to me Mrs. Somerville, whose bust I saw at Chantrey's?" The gentleman was Mr. Sopwith, of Newcastle-on-Tyne, a civil and mining engineer. He was distinguished for scientific knowledge, and had been in London to give information to a parliamentary committee. He travelled faster than we did, and when we arrived at Newcastle he was waiting to take us to his house, where we were hospitably received by Mrs. Sopwith. His conversation was highly interesting, and to him I was indebted for much information on mining generally, and on the mineral wealth of Great Britain, while writing on Physical Geography. Many years after he and Mrs. Sopwith came and saw me at Naples, which gave me much pleasure. He was unlike any other traveller I ever met with, so profound and original were his observations on all he saw.

* * * * *

On coming home I found that I had made an error in the first edition of the "Physical Sciences," in giving 365 days 6 hours as the length of the civil year of the ancient Egyptians. My friend Mr. Hallam, the historian, wrote to me, proving from history and epochs of the chronology of the ancient Egyptians, that their civil year was only 365 days. I was grateful to that great and amiable man for copies of all his works while he was alive, and I am obliged to his daughter for an excellent likeness of him, now that he is no more.

FROM HENRY HALLAM, ESQ., TO MRS. SOMERVILLE.

WIMPOLE STREET, *March 12th*, 1835.

MY DEAR MADAM,

As you will probably soon be called upon for another edition of your excellent work on the "Connexion of the Physical Sciences," I think you will excuse the liberty I take in mentioning to you one passage which seems to have escaped your attention in so arduous a labour. It is in page 104, where you have this sentence:—

"The Egyptians estimated the year at 365d. 6h., by which they lost one year in every 14,601, their Sothiac period. They determined the length of their year by the heliacal rising of Sirius, 2782 years before the Christian era, which is the earliest epoch of Egyptian chronology."

The Egyptian civil year was of 365 days only, as we find in Herodotus, and I apprehend there is no dispute about it. The Sothiac period, or that cycle in which the heliacal rising of Sirius passed the whole civil year, and took place again on the same day, was of 1461 years, not 14,601. If they had adopted a year of 365d. 6h., this period would have been more than three times 14,601; the excess of the sidereal year above that being only 9′ 9″, which will not amount to a day in less than about 125 years.

I do not see how the heliacal rising of Sirius in any one year could help them to determine its length. By comparing two successive years they could of course have got at a sidereal year; but this is what they did not do; hence the irregularity which produced the canicular cycle.

The commencement of that cycle is placed by ancient chronologers in 1322 A.C. It seems not correct to call 2782 A.C. "the earliest epoch of Egyptian chronology," for we have none of their chronology nearly so old, and in fact no chronology, properly so called, has yet been made out by our Egyptian researches. It is indeed certain that, if the reckoning by heliacal risings of Sirius did not begin in 1322, we must go nearly 1460 years back for its origin; since it must have been adopted when that event preceded only for a short time the annual inundation of the Nile. But, according to some, the year 1322 A.C. fell during the reign of Sesostris, to whom Herodotus ascribes several regulations connected with the rising of the Nile. Certainly, 2782 A.C. is a more remote era than we are hitherto warranted to assume for any astronomical observation.

Believe me, dear Mrs. Somerville,

Very truly yours,

HENRY HALLAM.

I refer you to Montucla, if you have any doubt about the Egyptian year being of 365 days without bissextile of any kind.

I had sent a copy of the "Mechanism of the Heavens" to M. Poisson soon after it was published, and I had received a letter from him dated 30th May, 1832, advising me to complete the work by writing a volume on the form and rotation of the earth and planets. Being again strongly advised to do so while in Paris, I now began the work, and, in consequence, I was led into a correspondence with Mr. Ivory, who had written on the subject, and also with Mr. Francis Baily, on the density and compression of the earth. My work was extensive, for it comprised the analytical attraction of spheroids, the form and rotation of the earth, the tides of the ocean and atmosphere, and small undulations.

When this was finished, I had nothing to do, and as I preferred analysis to all other subjects, I wrote a work of 246 pages on curves and surfaces of the second and higher orders. While writing this, *con amore*, a new edition of the "Physical Sciences" was much needed, so I put on high pressure and worked at both. Had these two manuscripts been published at that time, they might have been of use; I do not remember why they were laid aside, and forgotten till I found them years afterwards among my papers. Long after the time I am writing about, while at Naples, I amused myself by repairing the time-worn parts of these manuscripts, and was surprised to find that in my eighty-ninth year I still retained facility in the "Calculus."

The second edition of the "Physical Sciences" was dedicated to my dear friend, Sir John Herschel. It went through nine editions, and has been translated into German and Italian. The book went through various editions in the United States, to the honour, but not to the profit, of the author. However, the publisher obligingly sent me a copy. I must say that profit was never an object with me: I wrote because it was impossible for me to be idle.

I had the honour of presenting a copy of my book to the Duchess of Kent at a private audience. The Duchess and Princess Victoria were alone, and received me very graciously, and conversed for half an hour with me. As I mentioned before, I saw the young Princess crowned: youthful, almost child-like as she was, she went through the imposing ceremony with all the dignity of a Queen.

[A few letters from some of my mother's friends, written at this period, may prove of interest. They are chiefly written to thank her for copies of the Preliminary Dissertation or of the "Physical Sciences." One from Lord Brougham concerns my mother's estimate of the scientific merit of Dr. Young, for whom she had the sincerest admiration, and considered him one of the first philosophers and discoursers of the age.

FROM MISS EDGEWORTH TO MRS. SOMERVILLE.

EDGWORTHTOWN, *May* 31*st*, 1832.

MY DEAR MRS. SOMERVILLE,

There is one satisfaction at least in giving knowledge to the ignorant, to those who know their ignorance at least, that they are grateful and humble. You should have my grateful and humble thanks long ago for the favour—the honour—you did me by sending me that Preliminary Dissertation, in which there is so much knowledge, but that I really wished to read it over and over again at some intervals of time, and to have the pleasure of seeing my sister Harriet read it, before I should write to you. She has come to us, and has just been enjoying it, as I knew she would. For my part, I was long in the state of the boa constrictor after a full meal—and I am but just recovering the powers of motion. My mind was so distended by the magnitude, the immensity, of what you put into it! I am afraid that if you had been aware how ignorant I was you would not have sent me this dissertation, because you would have felt that you were throwing away much that I could not understand, and that could be better bestowed on scientific friends capable of judging of what they admire. I can only assure you that you have given me a great deal of pleasure; that you have enlarged my conception of the sublimity of the universe, beyond any ideas I had ever before been enabled to form.

The great simplicity of your manner of writing, I may say of your *mind*, which appears in your writing, particularly suits the scientific sublime—which would be destroyed by what is commonly called fine writing. You trust sufficiently to the natural interest of your subject, to the importance of the facts, the beauty of the whole, and the adaptation of the means to the ends, in every part of the immense whole. This reliance upon your reader's feeling along with you, was to me very gratifying. The ornaments of eloquence dressing out a sublime subject are just so many proofs either of bad taste in the orator, or of distrust and contempt of the taste of those whom he is trying thus to captivate.

I suppose nobody yet has completely *mastered* the tides, therefore I may well content myself with my inability to comprehend what relates to them. But instead o plaguing you with an endless enumeration of my difficulties, I had better tell you some of the passages which gave me, ignoramus as I am, peculiar pleasure . . . . . . I am afraid I shall transcribe your whole book if I go on to tell you all that has struck me, and you would not thank me for that—you, who have so little vanity, and so much to do better with your time than to read *my* ignorant admiration. But pray let me mention to you a few of the passages that amused my imagination particularly, viz., 1st, the inhabitant of Pallas *going round* his world—or who might go—in five or six hours in one of our steam carriages; 2nd, the moderate-sized man who would weigh two tons at the surface of the sun—and who would weigh only a few pounds at the surface of the four new planets, and would be so light as to find it impossible to stand from the excess of muscular force! I think a very entertaining dream might be made of a man's visit to the sun and planets—these ideas are all like dreamy feelings when one is a little feverish. I forgot to mention (page 58) a passage on the propagation of sound. It is a beautiful sentence, as well as a sublime idea, "so that at a very small height above the surface of the earth, the noise of the tempest ceases and the thunder is heard no more in those boundless regions, where the heavenly bodies accomplish their periods in eternal and sublime silence."

Excuse me in my trade of sentence-monger, and believe

me, dear Mrs. Somerville, truly your obliged and truly your affectionate friend,

MARIA EDGEWORTH.

I have persuaded your dear curly-headed friend, Harriet, to add her own observations; she sends her love to you; and I know you love her, otherwise I would not press her to write her own *say*.

FROM MISS JOANNA BAILLIE TO MRS. SOMERVILLE.

HAMPSTEAD, *February* 1*st*, 1832.

MY DEAR MRS. SOMERVILLE,

I am now, thank God! recovered from a very heavy disease, but still very weak. I will not, however, delay any longer my grateful acknowledgments for your very flattering gift of your Preliminary Dissertation. Indeed, I feel myself greatly honoured by receiving such a mark of regard from one who has done more to remove the light estimation in which the capacity of women is too often held, than all that has been accomplished by the whole sisterhood of poetical damsels and novel-writing authors. I could say much more on this subject were I to follow my own feelings; but I am still so weak that writing is a trouble to me, and I have nearly done all that I am able.

God bless and prosper you!

Yours gratefully and truly,

J. BAILLIE.

FROM MISS BERRY TO MRS. SOMERVILLE.

BELLEVUE, 18*th September*, 1834.

MY DEAR MRS. SOMERVILLE,

I have just finished reading your book, which has entertained me extremely, and at the same time, I hope, improved my moral character in the Christian virtue of humility. These must appear to you such *odd* results—so little like those produced on the great majority of your readers, that you must allow me to explain them to you. Humbled, I must be, by finding my own intellect unequal to following, beyond a first step, the explanations by which you seek to make easy to comprehension the marvellous phenomena of the universe—humbled, by feeling the intellectual difference between you and me, placing you as much above me in the scale of reasoning beings, as I am above my dog. Still I rejoice with humility at feeling myself, in that order of understandings which, although utterly incapable of following the chain of your reasonings, calculations, and inductions—utterly deprived of the powers necessary *sic itur ad astra*—am yet informed, enlightened, and entertained with the series of sublime truths to which you conduct me.

In some foggy morning of November, I shall drive out to you at Chelsea and surprise you with my ignorance of science, by asking you to explain to me some things which you will *wonder any one* can have so long existed without knowing. In the mean time, I wish you could read in any combination of the stars the probability of our often having such a season as this, of uninterrupted summer since April last, and when last week it was sobering into autumn, has now returned to enter summer again. The thermometer was at 83° in the shade yesterday, and to-day promises to be as much. We are delighted with our two months' residence at this place, which we shall see with regret draw towards a close the end of this month. October we mean to spend at Paris, before we return to the *nebulosities* of London. During my residence in Paris, before we came here, I never had the good luck to meet with your friend M. Arago; had I not been reading your book, I should have begged you to give me a letter for him. But as it is, and as my stay at Paris will now be so short, I shall content myself with looking up at a respectful distance to all your great fixed stars of science, excepting always yourself, dear Mrs. Somerville. No "disturbing influence" will, I hope, ever throw me out of the orbit of *your* intimacy and friendship, whose value, believe me, is most duly and accurately calculated by your ignorant but very affectionate friend,

M. Berry.

FROM LORD BROUGHAM TO MRS. SOMERVILLE.

1834.

My dear Mrs. Somerville,

Many thanks for the sheets, which I have read with equal pleasure and instruction as those I formerly had from you. One or two things I could have troubled you with, but they are of little moment. I shall note them. The only one that is at all material relates to the way you mention Dr. Young—not that I object to the word "illustrious," or as applied to him. But as you don't give it to one considerably more so, it looks either as if you overrated him, or underrated Davy, or (which I suppose to be the truth) as if you felt Young had not had his due share of honour, and desired to make it up to his memory. Observe I give him a very high place—but Davy's discoveries are both of more unquestioned originality and more undoubtedly true—perhaps I should say, more brought to a close. The alkalis and the principle of the safety lamp are concluded and fixed, the undulation is in progress, and somewhat uncertain as to how and where it may end. You will please to observe that I reckon both those capital discoveries of Davy the fruit of inquiry, and not at all of chance—for, as to the lamp, it is plain; and as to the metals, if you look at the inquiries that immediately preceded, you will see he was thereby led to the alkalis. Indeed, I well remember saying, when I read them, "He will analyse lime and barytes." I am quite ready to admit his extreme folly in some things, but that is nothing to the present purpose.

Yours,
H. B.
(*Henry Brougham.*)

FROM MRS. MARCET TO MRS. SOMERVILLE.

GENEVA, *6th April*, 1834.

DEAR MRS. SOMERVILLE,

I am desired by Professor Prevost to inform you that you were elected an honorary member of the Société de Physique et d'Histoire Naturelle de Genève on the 3rd April, and that a diploma will be forwarded to you by the earliest opportunity. After all the honours you have received, this little feather is hardly worthy of waving in your plume, but I am glad that Geneva should know how to appreciate your merit. You receive great honours, my dear friend, but that which you confer on our sex is still greater, for with talents and acquirements of masculine magnitude you unite the most sensitive and retiring modesty of the female sex; indeed, I know not any woman, perhaps I might say, any human being, who would support so much applause without feeling the weakness of vanity. Forgive me for allowing my pen to run away with this undisguised praise, it looks so much like compliment, but I assure you it comes straight from the heart, and you *must* know that it is fully deserved. . . . . I know not whether you have heard of the death of Professor de la Rive (the father); it was an unexpected blow, which has fallen heavily on all his family. It is indeed a great loss to Geneva, both as a man of science and a most excellent citizen.

M. Rossi* has left us to occupy the chair of political economy of the late M. Say, at Paris; his absence is sadly felt, and it is in vain to look around for any one capable of replacing him. . . . .

Yours affectionately,

J. Marcet.

FROM ADMIRAL W. H. SMYTH TO MRS. SOMERVILLE.

Crescent, Bedford, *October 3rd*, 1835.

My dear Madam,

As an opportunity offers of sending a note to town, I beg to mention that I have somewhat impatiently waited for some appearance of settled weather, in order

* M. Pellegrino Rossi, afterwards Minister of France at Rome, then Prime Minister to Pius the Ninth; murdered in 1848 on the steps of the Cancelleria, at Rome.

to press your coming here to inspect Halley's comet, before it should have become visible to the unassisted eye. That unerring monitor, however, the barometer, held forth no hope, and the ceaseless traveller is already an object of conspicuous distinction without artificial aid, except, perhaps, to most eyes an opera-glass, magnifying three or four times, will be found a pleasant addition. It is now gliding along with wonderful celerity, and the nucleus is very bright. It is accompanied with a great luminosity, and the nucleus has changed its position therein; that is, on the 29th August, the nucleus was like a minute star near the centre of the nebulous envelope; on the 2nd September it appeared in the *n. f.* quarter, and latterly it has been in the *s. f.*

How remarkable that the month of August this year should rattle Halley's name throughout the globe, in identity with an astonishing scientific triumph, and that in the selfsame month the letters of Flamsteed should have appeared! How I wish some one would give us a life of Newton, with all the interesting documents that exist of his labours! Till such appears, Flamsteed's statements, though bearing strong internal evidence of truth, are *ex-parte*, and it is evident his anxiety made him prone to impute motives which he could not prove. The book is painfully interesting, but except in all that relates to the personal character of Flamsteed, I could almost have wished the documents had been destroyed. People of judgment well know that men without faults are monsters, but vulgar minds delight in seeing the standard of human excellence lowered.

Dear Madam,

Yours faithfully,

W. H. Smyth.

We were deprived of the society of Sir John and Lady Herschel for four years, because Sir John took his telescope and other instruments to the Cape of Good Hope, where he went, accompanied by his family, for the purpose of observing the celestial phenomena of the southern hemisphere. There are more than 6,000 double stars in the northern hemisphere, in a large proportion of which the angle of position and distance between the two stars have been measured, and Sir John determined, in the same manner, 1081 in the southern hemisphere, and I believe many additions have been made to them since that time. In many of these one star revolves rapidly round the other. The elliptical orbits and periodical times of sixteen or seventeen of these stellar systems have been determined. In Gamma Virginis the two stars are nearly of the same magnitude, and were so far apart in the middle of the last century that they were considered to be quite independent of each other. Since then they have been gradually approaching one another, till, in March, 1836, I had a letter from Admiral Smyth, informing me that he had seen one of the stars eclipse the other, from his observatory at Bedford.

FROM ADMIRAL SMYTH TO MRS. SOMERVILLE.

CRESCENT, BEDFORD, *March 26th*, 1836.

MY DEAR MADAM,

Knowing the great interest you take in sidereal astronomy, of which so little is yet known, I trust it will not be an intrusion to tell you of a new, extraordinary, and very unexpected fact, in the complete occultation of one "fixed" star by another, under circumstances which admit of no possible doubt or equivocation.

You are aware that I have been measuring the position and distance of the two stars γ^1 and γ^2 Virginis, which are both nearly of similar magnitudes, and also, that they have approximated to each other very rapidly. They were very close last year, and I expected to find they had crossed each other at this apparition, but to my surprise I find they have become a fair round disc, which my highest powers will not elongate—in fact, *a single star!* I shall watch with no little interest for the reappearance of the second γ.

My dear madam,

Your truly obliged servant,

W. H. SMYTH.

This eclipse was also seen by Sir John Herschel at the Cape of Good Hope, as well as by many astronomers in Europe provided with instruments of great optical power. In 1782 Sir William Herschel saw one of the stars of Zeta Herculis eclipse the other.

In the "Connexion of the Physical Sciences" I have given an abridged account of Sir John Herschel's most remarkable discoveries in the southern hemisphere; but I may mention here that he determined the position and made accurate drawings of all the nebulæ that were distinctly visible in his 20 ft. telescope. The work he published will be a standard for ascertaining the changes that may take place in these mysterious objects for ages to come. Sir William Herschel had determined the places of 2,500 nebulæ in the northern hemisphere; they were examined by his son, and drawings made of some of the most remarkable, but when these nebulæ were viewed through Lord Rosse's telescope, they presented a very different appearance, showing that the apparent form of the nebulæ depends upon the space-penetrating power of the telescope, a circumstance of vital importance in observing the changes which time may produce on these wonderful objects.

[Long afterwards Lord Rosse wrote in reply to some questions which my mother had addressed to him on this subject:—

FROM THE EARL OF ROSSE TO MRS. SOMERVILLE.

CASTLE, PARSONSTOWN, *June 12th*, 1844.

DEAR MRS. SOMERVILLE,

I have very reluctantly postponed so long replying to your inquiries respecting the telescope, but there were some points upon which I was anxious to be enabled to speak more precisely. The instrument we are now using is 3 feet aperture, and 27 feet focus, and in the greater proportion of the nebulæ which have been observed with it some new details have been brought out. Perhaps the most interesting general result is that, as far as we have gone, increasing optical power has enlarged the list of clusters, by diminishing that of the nebulæ properly so-called. Such has always been the case since the nebulæ have been observed with telescopes, and although it would be unsafe to draw the inference, it is impossible not to feel some expectation that with sufficient optical power the nebulæ would all be reduced into clusters. Perhaps the two of the most remarkable of the resolved nebulæ are Fig. 26 and Fig. 55. In several of the planetary nebulæ we have discovered a star or bright point in the centre, and a filamentous edge, which is just the appearance which a cluster with a highly condensed centre would present in a small instrument. For instance, Figs. 47 and 32. We have also found that many of the nebulæ have not a symmetrical form, as they appear to have in inferior instruments; for instance, Fig. 81 is a cluster with long resolvable filaments from its southern extremity, and Fig. 85 is an oblong cluster with a bright centre. Fig. 45 is an annular nebula, like Herschel's drawing of the annular nebula in Lyra. I have sent drawings of a few of these objects to the Royal Society, they were forwarded a few days ago. We have upon the whole as yet observed but little with the telescope of 3 feet aperture. You recollect Herschel said that it was a good observing year, in which there were 100 hours fit for observing, and of the average of our hours I have not employed above 30. We have been for the last two years engaged in constructing a telescope of 6 feet aperture and 52 feet focus, and it would have been impossible to have bestowed the necessary attention upon it had we made a business of observing. That instrument is nearly finished, and I hope it will effect something for astronomy. The unequal refraction of the atmosphere will limit its powers, but how far remains to be ascertained. . . . . Lady Rosse joins me in very kind remembrances and believe me to be,

Dear Mrs. Somerville,

Yours very truly and ever,

Rosse.

[Sir John Herschel wrote to my father from the Cape:—

FROM SIR JOHN HERSCHEL TO MR. SOMERVILLE.

Feldhausen, near Wynberg, C. G. H., *July 17th*, 1830.

My dear Somerville,

Since our arrival here, I have, I know in many instances, maintained or established the character of a bad correspondent; and really it is not an inconvenient character to have established. Only, in your case, I should be very sorry to appear in that, or any other negligent or naughty light; but you, I know, will allow for the circumstances which have occasioned my silence. Meanwhile, I am not sorry that the execution of an intention I had more than once formed should have been deferred, till we read in the papers of the well-judged and highly creditable notice (creditable I mean to the government *pro tempore*) which His Majesty has been pleased to take of Mrs. Somerville's elaborate works. Although the Royal notice is not quite so swift as the lightning in the selection of its objects, it agrees with it in this, that it is attracted by the loftiest; and though what she has performed may seem so natural and easy to herself, that she may blush to find it fame; all the rest of the world will agree with me in rejoicing that merit of that kind is felt and recognised at length in the high places of the earth. This, and the honourable mention of Airy by men of both parties in the House of Commons about the same time, are things that seem to mark the progress of the age we live in; and I give Peel credit for his tact in perceiving this mode of making a favourable impression on the public mind.

We are all going on very comfortably, and continue to like the Cape as a place of (temporary) residence as much or more than at first. The climate is so very delicious. . . . . . The stars are most propitious, and, astronomically speaking, I can now declare the climate to be most excellent. Night after night, for weeks and months, with hardly an interruption, of *perfect* astronomical weather, discs of stars reduced almost to points, and tranquilly gliding across the field of your telescope. It is really a treat, such as occurs once or perhaps twice a year in England—hardly more. I had almost forgotten that by a recent vote of the Astronomical Society I can now claim Mrs. Somerville as a *colleague*. Pray make my compliments to her in that capacity, and tell her that I hope to meet her there at some future session. . . .

Yours very faithfully,

H. W. Herschel.

To William Somerville, Esq.

Spectrum analysis has shown that there is a vast quantity of self-luminous gaseous matter in space, incapable of being reduced into stars, however powerful the telescope through which it is observed. Hence the old opinion once more prevails, that this is the matter of which the sun and stellar systems have been formed, and that other stellar systems are being formed by slow, continuous condensation. The principal constituents of this matter are, the terrestrial gases, hydrogen, and nitrogen. The yellow stars, like the sun, contain terrestrial matter. The nebulous and stellar constituents were chiefly discovered by Dr. Huggins.

Somerville and I were always made welcome by Sir James South, and at Camden Hill I learnt the method of observing, and sometimes made observations myself on the double stars and binary systems, which, worthless as they were, enabled me to describe better what others had done. One forenoon Somerville and I went to pay a visit to Lady South. Sir James, who was present, said, "Come to the observatory, and measure the distance of Mercury from the sun; for they are in close approximation, and I wish to see what kind of observation you will make." It was erroneous, as might have been expected; but when I took the mean of several observations, it differed but little from that which Sir James South had made; and here I learnt practically the importance of taking the mean of approximate quantities.

* * * * *

Dr. Wollaston, Dr. Young, and the Katers died before I became an author; Lord Brougham was one of the last of my scientific contemporaries, all the rest were younger than myself, and with this younger set, as with their predecessors, we had most agreeable and constant intercourse. Although we lived so much in scientific society we had all along been on the most friendly and intimate terms with the literary society of the day, such as Hallam, Milman, Moore, Malthus, &c., &c. The highly intellectual conversation of these was enlivened by the brilliant wit of my early friend, Sydney Smith, who was loved and admired by every one. His daughter married our friend Sir Henry Holland, the distinguished physician, well known for his eminent literary and scientific acquirements as well as for his refined taste.

No house in London was more hospitable and agreeable than that of the late Mr. John Murray, in Albemarle Street. His dinner parties were brilliant, with all the poets and literary characters of the day, and Mr. Murray himself was gentlemanly, full of information, and kept up the conversation with spirit. He generously published the "Mechanism of the Heavens" at his own risk, which, from its analytical character, could only be read by mathematicians.

Besides those I have mentioned we had a numerous acquaintance who were neither learned nor scientific; and at concerts at some of their houses I enjoyed much hearing the great artists of the day, such as Pasta, Malibran, Grisi, Rubini, &c., &c. We knew Lucien Buonaparte, who gave me a copy of his poems, which were a failure.

I had become acquainted with Madame de Montalembert, who was an Englishwoman, and was mother of the celebrated Comte; she was very eccentric, and at that time was an Ultra-Protestant. One day she came to ask me to go and drive in the Park with her, and afterwards dine at her house, saying, "We shall all be in high dresses." So I accepted, and on entering the drawing-room, found a bishop and several clergymen, Lady Olivia Sparrow, and some other ladies, all in high black satin dresses and white lace caps, precisely the dress I wore, and I thought it a curious coincidence. The party was lively enough, and agreeable, but the conversation was in a style I had never heard before—in fact, it affected the phraseology of the Bible. We all went after dinner to a sort of meeting at Exeter Hall, I quite forget for what purpose, but our party was on a kind of raised platform. I mentioned this to a friend afterwards, and the curious circumstance of our all being dressed alike. "Do you not know," she said, "that dress is assumed as a distinctive mark of the Evangelical party! So you were a wolf in sheep's clothing!"

I had been acquainted with the Miss Berrys at Raith, when visiting their cousins, Mr. and Mrs. Ferguson. Mary, the eldest, was a handsome, accomplished woman, who from her youth had lived in the most distinguished society, both at home and abroad. She published a "Comparative View of Social Life in France and England," which was well received by the public. She was a Latin scholar, spoke and wrote French fluently, yet with all these advantages, the consciousness that she might have done something better, had female education been less frivolous, gave her a characteristic melancholy which lasted through life. She did not talk much herself, but she had the tact to lead conversation. She and her sister received every evening a select society in their small house in Curzon Street. Besides any distinguished foreigners who happened to be in London, among their habitual guests were my friend, Lady Charlotte Lindsay, always witty and agreeable, the brilliant and beautiful Sheridans, Lady Theresa Lister, afterwards Lady Theresa Lewis, who edited Miss Berry's "Memoirs," Lord Lansdowne, and many others. Lady Davy came occasionally, and the Miss Fanshaws, who were highly accomplished, and good artists, besides Miss Catherine Fanshaw wrote clever *vers de société*, such as a charade on the letter H, and, if I am not mistaken, "The Butterfly's Ball," &c. I visited these ladies, but their manners were so cold and formal that, though I admired their talents, I never became intimate with them. On the contrary, like everyone else, I loved Mary Berry, she was so warm-hearted and kind. When London began to fill, and the season was at its height, the Miss Berrys used to retire to a pretty villa at Twickenham, where they received their friends to luncheon, and strawberries and cream, and very delightful these visits were in fine spring weather. I recollect once, after dining there, to have been foruntate enough to give a place in my carriage to

Lord Macaulay, and those who remember his charming and brilliant conversation will understand how short the drive to London appeared.

We sometimes went to see Miss Lydia White, who received every evening; she was clever, witty, and very free in her conversation. On one occasion the party consisted, besides ourselves, of the Misses Berry, Lady Davy; the three poets, Rogers, William Spencer, and Campbell; Sir James Macintosh, and Lord Dudley. Rogers, who was a bitter satirist and hated Lord Dudley, had written the following epigram:—

Ward has no heart, 'tis said; but I deny it.
He has a heart, and gets his speeches by it.

I had never heard of this epigram, and on coming away Lord Dudley said, "You are going home to sleep and I to work." I answered, "Oh! you are going to prepare your speech for to-morrow." My appropriate remark raised an universal laugh.

* * * * *

Mr. Bowditch, of Boston, U. S., who died in 1838, left among other works a "Commentary on La Place's Mécanique Céleste" in four volumes. While busily occupied in bringing out an edition of the "Physical Sciences," I received a letter from his son, Mr. H. Bowditch, requesting me to write an elaborate review of that work, which would be published in Boston along with the biography of his father, written by Mr. Young, who sent me a copy of it. Though highly sensible of the honour, I declined to undertake so formidable a work, fearing that I should not do justice to the memory of so great a man.

I have always been in communication with some of the most distinguished men of the United States. Washington Irving frequently came to see me when he was in London; he was as agreeable in conversation as he was distinguished as an author. No one could be more amiable than Admiral Wilkes, of the U. S. navy: he had all the frankness of a sailor. We saw a good deal of him when he was in London, and I had a long letter from him, giving me an account of his fleet, his plan for circumnavigation, &c. &c. I never had the good fortune to become personally acquainted with Captain Maury, of the U. S. navy, author of that fascinating book, the "Physical Geography of the Sea," but I am indebted to him for a copy of that work, and of his valuable charts. Mr. Dana, who is an honour to his country, sent me copies of his works, to which I have had occasion frequently to refer as acknowledged authority on many branches of natural history. I should be ungrateful if I did not acknowledge the kindness I received from the Silliman family, who informed me of any scientific discovery in the United States, and sent me a copy of their Journal when it contained anything which might interest me. I was elected an honorary member of the Geographical and Statistical Society of New York, U. S. on the 15th May, 1857, and on the 15th October, 1869, I was elected a member of the American Philosophical Society at Philadelphia, for Promoting Useful Knowledge. I shall ever be most grateful for these honours.

While living in Florence, many years after, an American friend invited me to an evening party to meet an American authoress who wished particularly to make my acquaintance. I accordingly went there on the evening in question, and my friends, after receiving me with their accustomed cordiality, presented me to the lady, and placed me beside her to give me an opportunity of conversing with her. I addressed her several times, and made various attempts to enter into conversation, but only received very dry answers in reply. At last she fairly turned her back upon me, and became engrossed with a lady who sat on her other side, upon which I got up and left her and never saw her again. A very different person in every respect was present that evening, as much distinguished by her high mental qualities and poetical genius as by her modesty and simplicity. I allude to our greatest British poetess, Mrs. Browning, who at that time resided in Florence, except when the delicacy of her health obliged her to go to Rome. I think there is no other instance of husband and wife both poets, and both distinguished in their different lines. I can imagine no happier or more fascinating life than theirs; two kindred spirits united in the highest and noblest aspirations. Unfortunately her life was a short one; in the full bloom of her intellect her frail health gave way, and she died leaving a noble record of genius to future ages, and a sweet memory to those who were her contemporaries. The Florentines, who, like all Italians, greatly appreciate genius, whether native or foreign, have placed a commemorative tablet on Casa Guidi, the house Mrs. Browning inhabited.

I was extremely delighted last spring in being honoured by a visit from Longfellow, that most genial poet. It is not always the case that the general appearance of a distinguished person answers to one's ideal of what he ought to be—in this respect Longfellow far surpasses expectation. I was as much charmed with his winning manner and conversation as by his calm, grand features and the expression of his intellectual countenance.

The Barons Fairfax, as I mentioned already, had

long been members of the Republic of the United States, and Washington's mother belonged to this family. During the war of Independence, while my father, then Lieutenant Fairfax, was on board a man-of-war on the American station, he received a letter from General Washington claiming him as a relation, and inviting him to pay him a visit, saying, he did not think that war should interfere with the courtesies of private life. Party spirit ran so high at that time that my father was reprimanded for being in correspondence with the enemy. I mentioned to my friend, the Rev. Dr. Tuckerman, of the United States, how much I regretted that so precious a letter had been lost, and he most kindly on going home sent me an autograph letter of General Washington.

FROM THE REV. JOSEPH TUCKERMAN TO MRS. SOMERVILLE.

BOSTON, *August 28th*, 1834.

MY DEAR MADAM,

I have very great pleasure in sending to you an autograph letter of your and our glorious Washington. I obtained it from Mr. Sparks, who had the gratification of seeing you when he was in England, and who told me when I applied to him for it, that there is no one in the world to whom he would be so glad to give it. It is beyond comparison the best and almost the only remaining one at his disposal among the "Washington" papers.

I am again in my family and in the field of my ministry.

But very dear to me are my associations with scenes and friends in England; and most glad should I be if I could renew that intercourse with yourself, and with the intellect and virtue around you, to which I have been indebted for great happiness, and which, I hope, has done something to qualify me for a more efficient service. Will you please to present my very sincere respects to your husband, and to recall me to the kind remembrance of your children. With the highest respect and regard, allow me to call myself,

Your friend,

JOSEPH TUCKERMAN.

I think it must have been on returning from the American station, or may be later in the career of my father's life, that a circumstance occurred which distressed him exceedingly. Highway robberies were common on all the roads in the vicinity of London, but no violence was offered. My father was travelling alone over Blackheath when the postilion was ordered to stop, a pistol presented at my father, and his purse demanded. My father at once recognised the voice as that of a shipmate, and exclaimed, "Good God! I know that voice! can it be young ——? I am dreadfully shocked; I have a hundred pounds which shall be yours—come into the carriage, and let me take you to London, where you will be safe." . . "No, no," the young man said, "I have associates whom I cannot leave—it is too late." . . . It was too late; he was arrested eventually and suffered. Years afterwards when by some accident my father mentioned this event, he was deeply affected, and never would tell the name of the young man who had been his messmate.

CHAPTER XIV.

ROME, NAPLES, AND COMO—BADEN—WINTER AT FLORENCE—SIENA—LETTER FROM LORD BROUGHAM—MR. MOUNTSTUART ELPHINSTONE—LIFE AT ROME—CAMPAGNA CATTLE.

[My mother was already meditating writing a book upon Physical Geography, and had begun to collect materials for it, when my father's long and dangerous illness obliged her to lay it aside for a time. My father was ordered to a warmer climate for the winter, and as soon as he was able to travel we proceeded to Rome. We were hardly settled when my mother, with her usual energy, set to work diligently, and began this book, which was not published for some time later, as it required much thought and research. She never allowed anything to interfere with her morning's work; after that was over she was delighted to join in any plan which had been formed for the afternoon's amusement, and enjoyed herself thoroughly, whether in visiting antiquities and galleries, excursions in the neighbourhood, or else going with a friend to paint on the Campagna. My mother was extremely fond of Rome, and often said no place had ever suited her so well. Independently of the picturesque beauty of the place, which, to such a lover of nature, was sufficient in itself, there was a very pleasant society during many seasons we spent there. The visitors were far less numerous than they are now, but on that very account there was more sociability and intimacy, and scarcely an evening passed without our meeting. The artists residing at Rome, too, were a most delightful addition to society. Some of them became our very dear friends. My mother remarks :—

WE took lodgings at Rome, and as soon as we were settled I resumed my work and wrote every morning till two o'clock, then went to some gallery, walked on the Pincio, dined at six, and in the evening either went out or received visits at home—the pleasantest way of seeing friends, as it does not interfere with one's occupations.

We once joined a party that was arranged to see the statues in the Vatican by torchlight, at which Lord Macaulay astonished us by his correct knowledge and learning as we passed through the gallery of inscriptions. To me this evening was memorable; on this occasion I first met with John Gibson, the sculptor, who afterwards became a dear and valued friend. He must have been a pupil of Canova's or Thorwaldsen's when Somerville and I were first at Rome. Now his fame was as great as that of either of his predecessors.

*   *   *   *   *

[In spring we went to Naples for a few weeks, and returned to Rome by the San Germano road, now so familiar to travellers, but then hardly ever frequented, as it was extremely unsafe on account of the brigands. We met with no adventures, although we often reached our night quarters long after sunset, for my mother sketched a great deal on the road. We travelled by vetturino and continued this delightful journey to Como. My mother was a perfect travelling companion, always cheerful and contented and interested in all she saw. I leave her to tell of our pleasant residence at Bellaggio in her own words:—

We remained only a short time at Florence, and then went for a month to Bellaggio, on the Lake of Como, at that time the most lonely village imaginable. We had neither letters, newspapers, nor any books, except the Bible, yet we liked it exceedingly. I did nothing but paint in the mornings, and Somerville sat by me. My daughters wandered about, and in the evening we went in a boat on the lake. Sometimes we made longer excursions. One day we went early to Menaggio, at the upper end of the lake. The day had been beautiful, but while at dinner we were startled by a loud peal of thunder. The boatmen desired us to embark without delay, as a storm was rising behind the mountains; it soon blew a gale, and the lake was a sheet of foam; we took shelter for a while at some place on the coast and set out again, thinking the storm had blown over, but it was soon worse than ever. We were in no small danger for two hours. The boatmen, terrified, threw themselves on their knees in prayer to the Madonna. Somerville seized the helm and lowered the sail and ordered them to rise, saying, the Madonna would help them if they helped themselves, and at last they returned to their duty. For a long time we remained perfectly silent, when one of our daughters said, "I have been thinking what a paragraph it will be in the newspapers, 'Drowned, during a sudden squall on the lake of Como, an English family named Somerville, father, mother and two daughters.'" The silence thus broken made us laugh, though our situation was serious enough, for when we landed the shore was crowded with people who had fully expected to see the boat go down. Twice after this we were overtaken by these squalls, which are very dangerous. I shall never forget the magnificence of the lightning and the grandeur of the thunder, which was echoed by the mountains during the storms on the Lake of Como.

We saw the fishermen spear the fish by torchlight, as they did on the Tweed. The fish were plenty and the water so clear that they were seen at a great depth. There are very large red-fleshed trout in the lake, and a small very delicious fish called *agoni*, caught in multitudes by fine silk nets, to which bells are attached on floats, that keep up a constant tinkling to let the fishermen know where to find their nets when floated away by the wind.

[We now crossed the Alps, by the St. Gothard, to Basle and Baden Baden, where we passed the summer, intending to return to England in autumn, but as soon as the rains began my father had so serious a return of his illness that my mother was much alarmed. When he was well enough to travel, we once more crossed the Alps, and reached Florence, where we remained for the winter. My mother resumed her work there.

Through the kindness of the Grand Duke, I was allowed to have books at home from his private library in the Pitti Palace, a favour only granted to the four Directors. This gave me courage to collect materials for my long neglected Physical Geography, still in embryo. As I took an interest in every branch of science I became acquainted with Professor Amici, whose microscopes were unrivalled at that time, and as he had made many remarkable microscopic discoveries in natural history, he took us to the Museum to see them magnified and modelled in wax. I had the honour of being elected a member of the Academy of Natural Science at Florence.

There were many agreeable people at Florence that winter and a good deal of gaiety. The Marchese Antinori presented Somerville and me to the Grand Duke, who had expressed a wish to know me. He received us very graciously, and conversed with us for more than an hour on general subjects. He afterwards wrote me a polite letter, accompanied by a work on the drainage of the Maremma, and gave directions about our being invited to a scientific meeting which was to be held at Pisa. We were presented to the Grand Duchess, who was very civil. We spent the summer at Siena, and had a cheerful airy apartment with a fine view of the hills of Santa Fiora, and with very pretty arabesques in fresco on the walls of all the rooms, some so very artistic that I made sketches of them. In these old cities many of the palaces and houses are decorated with that artistic taste which formerly prevailed to such an extent in Italy, and which has now yielded, here as elsewhere, to common-place modern furniture.

*　　*　　*　　*　　*

[While we were at Siena, my mother received the following letter from Lord Brougham, who was a frequent correspondent of hers, but whose letters are generally too exclusively mathematical for the general reader. My mother had described the curious horse-races which are held at Siena every three years, and other mediæval customs still prevalent.

FROM LORD BROUGHAM TO MRS. SOMERVILLE.

COLE HILL, KENT, *Sept. 28th,* 1840.

MY DEAR MRS. SOMERVILLE,

I am much obliged to you for your kind letter which let me know of your movements. I had not heard of them since I saw the Fergusons. . . . . . We have been here since parliament rose, as I am not yet at all equal to going to Brougham. My health is now quite restored; but I shall not soon—nor in all probability ever—recover the losses I have been afflicted with. I passed the greater part of last winter in Provence, expecting some relief from change of scene and from the fine climate; but I came back fully worse than when I went. In fact, I did wrong in struggling at first, which I did to be able to meet parliament in January last. If I had yielded at once, I would have been better. I hope and trust they sent you a book I published two years ago; I mean the "Dissertations," of which one is on the "Principia," and designed to try how far it may be taught to persons having but a very moderate stock of mathematics; also, if possible, to keep alive the *true taste* (as I reckon it) in mathematics, which modern analysis has a little broken in upon. Assuming you to have got the book, I must mention that there are some intolerable errors of the press left, such as . . . . . . Excuse my troubling you with these errata, and impute it to my wish that you should not suppose me to have written the nonsense which these pages seem to prove. By the way, it is a curious proof of university prejudice, that though the Cambridge men admit my analysis of the "Principia" to be unexceptionable, and to be well calculated for teaching the work, yet, *not being by a Cambridge man,* it cannot be used! They are far more liberal at Paris, where they only are waiting for my analysis of the second book; but I put off finishing it, as I do still more my account of the "Mécanique Céleste." The latter I have almost abandoned in despair after nearly finishing it; I find so much that cannot be explained elementarily, or anything near it. So that my account to be complete would be nearly as hard reading as yours, and not 1000th part as good . . . . I greatly envy you Siena; I never was there above a day, and always desired to stay longer. The language is, as you say, a real charm; but I was not aware of the preservation in which you describe the older manners to be. I fear I shall not be able to visit Provence, as I should have wished this winter . . . . . . but my plans are not quite fixed. The judicial business in Parliament and the Privy Council will also make my going abroad after January difficult. I don't write you any news, nor is there any but what you see in the papers. The Tory restoration approaches very steadily, tho' not very rapidly; and I only hope that the Whigs, having contrived to destroy the Liberal party in the country—I fear past all hope of recovery—may not have a war abroad also to mourn for. . . . . .

Believe me,

Yours ever,

H. Brougham.

On going to Rome I required a good many books for continuing my work on "Physical Geography," and had got "Transactions of the Geographical Society" and other works sent from London. The Hon. Mountstuart Elphinstone who was then at Rome, was an old acquaintance of ours. He was one of the most amiable men I ever met with, and quite won my heart one day at table when they were talking of the number of singing-birds that were eaten in Italy—nightingales, gold-finches, and robins—he called out, "What! robins! our household birds! I would as soon eat a child!" He was so kind as to write to the Directors of the East India Company requesting that I might have the use of the library and papers that were in the India House. This was readily granted me; and I had a letter in consequence from Mr. Wilson, the Orientalist, giving me a list of the works they had on the geography of Eastern Asia and the most recent travels in the Himalaya, Thibet, and China, with much useful information from himself. I was indebted to Sir Henry Pottinger, then at Rome, for information relating to Scinde, for he had been for some years British Envoy at Beloochistan. Thus provided, I went on with my work. We lived several winters in an apartment on the second floor of Palazzo Lepri, Via dei Condotti, where we passed many happy days. When we first lived in Via Condotti, the waste-pipes to carry off the rain-water from the roofs projected far into the street, and when there was a violent thunderstorm, one might have thought a waterspout had broken over Rome, the water poured in such cascades from the houses on each side of the street. On one occasion the rain continued in torrents for thirty-six hours, and the Tiber came down in heavy flood, inundating the Ghetto and all the low parts of the city; the water was six feet deep in the Pantheon. The people were driven out of their houses in the middle of the night and took refuge in the churches, and boats plied in the streets supplying the inhabitants with food, which they hauled up in baskets let down from the windows. The Campagna for miles was under water; it covered the Ponte Molle so that the courier could not pass; and seen from the Pincio it looked like an extensive lake. Much anxiety was felt for the people who lived in the farm houses now surrounded with water. Boats were sent to rescue them, and few lives were lost; but many animals perished. The flood did not subside till after three days, when it left everything covered with yellow mud; the loss of property was very great, and there was much misery for a long time.

Our house was in a very central position, and when not engaged I gladly received anyone who liked to come to us in the evening, and we had a most agreeable society, foreign and English, for we were not looked upon as strangers, and the English society was much better during the years we spent in Rome than it was afterwards.

I had an annual visit of an hour from the astronomer Padre Vico, and Padre Pianciani, Professor of Chemistry in the Collegio Romano. I was invited to see the Observatory; but as I had seen those of Greenwich and Paris, I did not think it worth while accepting the invitation, especially as it required an order from the Pope. I could easily have obtained leave, for we were presented to Gregory XVI. by the President of the Scotch Catholic College. The Pope received me with marked distinction; notwithstanding I was disgusted to see the President prostrate on the floor, kissing the Pope's foot as if he had been divine. I think it was about this time that I was elected an honorary associate of the Accademia Tiberiana.

I had very great delight in the Campagna of Rome; the fine range of Apennines bounding the plain, over which the fleeting shadows of the passing clouds fell, ever changing and always beautiful, whether viewed in the early morning, or in the glory of the setting sun, I was never tired of admiring; and whenever I drove out, preferred a country drive to the more fashionable Villa Borghese. One day Somerville and I and our daughters went to drive towards the Tavolata, on the road to Albano. We got out of the carriage, and went into a field, tempted by the wild flowers. On one side of this field ran the aqueduct, on the other a deep and wide ditch full of water. I had gone towards the aqueduct, leaving the others in the field. All at once we heard a loud shouting, when an enormous drove of the beautiful Campagna grey cattle with their wide-spreading horns came rushing wildly between us with their heads down and their tails erect, driven by men with long spears mounted on little spirited horses at full gallop. It was so sudden and so rapid, that only after it was over did we perceive the danger we had run. As there was no possible escape, there was nothing for it but standing still, which Somerville and my girls had presence of mind to do, and the drove dividing, rushed like a whirlwind to the right and left of them. The danger was not so much of being gored as of being run over by the excited and terrified animals, and round the walls of Rome places of refuge are provided for those who may be passing when the cattle are driven. Near where this occurred there is a house with the inscription "*Casa Dei Spiriti*"; but I do not think the Italians believe in either ghosts or witches; their chief superstition seems to be the "*Jettatura*," or evil eye, which they have inherited from the early Romans, and, I believe, Etruscans. They consider it a bad omen to meet a monk or priest on first going out in the morning. My daughters were engaged to ride with a large party, and the meet was at our house. A Roman, who happened to go out first, saw a friar, and rushed in again laughing, and waited till he was out of sight. Soon after they set off, this gentleman was thrown from his horse and ducked in a pool; so the "*Jettatura*" was fulfilled. But my daughters thought his bad seat on horseback enough to account for his fall without the Evil Eye.

CHAPTER XV.

ALBANO—POPULAR SINGING—LETTERS FROM MRS. SOMERVILLE—GIBSON—PERUGIA—COMET OF 1843—SUMMER AT VENICE—LETTERS FROM MRS. SOMERVILLE AND MISS JOANNA BAILLIE—ELECTED ASSOCIATE OF THE COLLEGE OF RESURGENTI AND R. I. ACADEMY OF SCIENCE AT AREZZO.

In spring we went to Albano, and lived in a villa, high up on the hill in a beautiful situation not far from the lake. The view was most extensive, commanding the whole of the Campagna as far as Terracina, &c. In this wide expanse we could see the thunderclouds forming and rising gradually over the sky before the storm, and I used to watch the vapour condensing into a cloud as it rose into the cool air. I never witnessed anything so violent as the storms we had about the equinox, when the weather broke up. Our house being high above the plain became enveloped in vapour till, at 3 p.m., we could scarcely see the olives which grew below our windows, and crash followed crash with no interval between the lightning and the thunder, so that we felt sure many places must have been struck; and we were not mistaken—trees, houses, and even cattle had been struck close to us. Somerville went to Florence to attend a scientific meeting, and wrote to us that the lightning there had stripped the gold leaf off the conductors on the powder magazine; a proof of their utility.

The sunsets were glorious, and I, fascinated by the gorgeous colouring, attempted to paint what Turner alone could have done justice to. I made studies, too, which were signal failures, of the noble ilex trees bordering the lake of Albano. Thus I wasted a great deal of time, I can hardly say in vain, from the pleasure I had in the lovely scenery. Somerville sat often by me with his book, while I painted from nature, or amused himself examining the geological structure of the country. Our life was a solitary one, except for the occasional visit from some friends who were at Frascati; but we never found it dull; besides, we made many expeditions on mules or donkeys to places in the neighbourhood. I was very much delighted with the flora on the Campagna and the Alban hills, which in spring and early summer are a perfect garden of flowers. Many plants we cultivate in England here grow wild in profusion, such as cyclamens, gum-cistus, both white and purple, many rare and beautiful orchideæ, the large flowering Spanish broom, perfuming the air all around, the tall, white-blossomed Mediterranean heath, and the myrtle. These and many others my girls used to bring in from their early morning walks. The flowers only lasted till the end of June, when the heat began, and the whole country became brown and parched; but scarcely had the autumnal rains commenced, when, like magic, the whole country broke out once more into verdure, and myriads of cyclamens covered the ground. Nightingales abounded in the woods, singing both by night and by day; and one bright moonlight night my daughters, who slept with their window open, were startled from their sleep by the hooting of one of those beautiful birds, the great-eared owl—"le grand duc" of Buffon—which had settled on the railing of their balcony. We constantly came across snakes, generally harmless ones; but there were a good many vipers, and once, when Somerville and my daughters, with Mr. Cromek, the artist, had gone from Genzano to Nettuno for a couple of days, a small asp which was crawling among the bent-grass on the sea-shore, darted at one of the girls, who had irritated it by touching it with her parasol. By the natives they are much dreaded, both on this coast and in the pine forest of Ravenna, where the cattle are said to be occasionally poisoned by their bite.

* * * * *

We had been acquainted with the Rev. Dr., afterwards Cardinal Wiseman at Rome. He was head of a college of young men educating for the Catholic Church, who had their "villeggiatura" at Monte Porzio. We spent a day with him there, and visited Tusculum ; another day we went to Lariccia, where there is a palace and park belonging to the Chigi family in a most picturesque but dilapidated state. We went also to Genzano, Rocca del Papa, and occasionally to visit friends at Frascati. There was a stone threshing-floor behind our house. During the vintage we had it nicely swept and lighted with torches, and the grape gatherers came and danced till long after midnight, to the great amusement of my daughters, who joined in the dance, which was the Saltarello, a variety of the Tarantella. They danced to the beating of tambourines. Italy is the country of music, especially of melody, and the popular airs, especially the Neapolitan, are extremely beautiful and melodious ; yet it is a fact, that the singing of the peasantry, particularly in the Roman and Neapolitan provinces, is most disagreeable and discordant. It is not melody at all, but a kind of wild chant, meandering through minor tones, without rhythm of any sort or apparent rule, and my daughters say it is very difficult to note down ; yet there is some kind of method and similarity in it as one hears it shouted out at the loudest pitch of the voice, the last note dwelt upon and drawn out to an immeasurable length. The words are frequently improvised by the singers, who answer one another from a distance, as they work in the fields. I have been told this style of chanting—singing it can hardly be called—has been handed down from the most ancient times, and it is said, in the southern provinces, to have descended from the early Greek colonists. The ancient Greeks are supposed to have chanted their poetry to music, as do the Italian improvisatori at the present day. In Tuscany, the words of the songs are often extremely poetical and graceful. Frequently, these verses, called "stornelli" and "rispetti," are composed by the peasants themselves, women as well as men ; the language is the purest and most classical Italian, such as is spoken at the present day in the provinces of Siena, Pistoja, &c., very much less corrupted by foreign idioms or adaptations than what is spoken, even by cultivated persons, in Florence itself. The picturesque costumes so universal when I first came to Italy, in 1817, had fallen very much into disuse when, at a much later period, we resided in Rome, and now they are rarely seen.

We hired a handsome peasant girl from Albano as housemaid, who was much admired by our English friends in her scarlet cloth bodice, trimmed with gold lace, and the silver spadone, or bodkin, fastening her plaits of dark hair; but she very soon exchanged her picturesque costume for a bonnet, etc., in which she looked clumsy and commonplace.

[The following are extracts from letters written from Albano by my mother:—

FROM MRS. SOMERVILLE TO HER SON W. GREIG, ESQ.

ALBANO, *16th June*, 1841.

I was thankful to hear, my dearest Woronzow, from your last letter that Agnes is recovering so well. . . . . We are very much pleased with our residence at Albano; the house, with its high sounding name of "Villa," is more like a farmhouse, with brick floors and no carpets, and a few chairs and tables, but the situation is divine. We are near the top of the hill, about half-a-mile above Albano, and have the most magnificent view in every direction, and such a variety of delightful walks, that we take a new one every evening. For painting it is perfect; every step is a picture. At present we have no one near, and lead the life of hermits; but our friends have loaded us with books, and with drawing, painting, music, and writing, we never have a moment idle. Almost every one has left Rome; but the English have all gone elsewhere, as they are not so easily pleased with a house as we are. The only gay thing we have done was a donkey ride yesterday to the top of Monte Cavo, and back by the lake of Nemi. . . . . . .

FROM MRS. SOMERVILLE TO WORONZOW GREIG, ESQ.

ALBANO, 29*th August*, 1841.

I dare say you think it very long since you have heard from me, my dearest Woronzow, but the truth is, I have been writing so hard, that after I had finished my day's work, I was fit for nothing but idleness. The reason of my hurry is, that the scientific meeting takes place at Florence on the 15th of September, and as I think it probable that some of our English philosophers will come to it, I hope to have a safe opportunity of sending home some MS. which it has cost me hard work to get ready, as I have undertaken a book more fit for the combination of a Society than for a single hand to accomplish. Lord Brougham was most kind when at Rome, and took so great an interest in it, that he has undertaken to read it over, and give me his opinion and criticism, which will be very valuable, as I know no one who is a better judge of these matters. He will send it to Mr. Murray, and you had better consult with him about it, whether he thinks it will succeed or not. Both William and Martha like what I have done; but I am very nervous about it, and wish you would read it if you have time. . . . . . . We have been extremely quiet all the summer; we have no neighbours, so that we amuse ourselves with our occupations. I get up between six and seven, breakfast at eight, and write till three, when we dine; after dinner, I write again till near six, when we go out and take a long walk; come home to tea at nine, and go to bed at eleven: the same thing day after day, so you cannot expect a very amusing letter. . . . . . . I have another commission I wish you would do for me; it is to inquire what discoveries Captain Ross has made at the South Pole. I saw a very interesting account in "Galignani" of what they have done, but cannot trust to a newspaper account so as to quote it.

A new edition of my "Physical Sciences" was required, so the "Physical Geography" was laid aside for the present. On returning to Rome, we resumed our usual life, and continued to receive our friends in the evening without ceremony. There was generally a merry party round the tea table in a corner of the room. I cannot omit mentioning one of the most charming and intellectual of our friends, Don Michelangelo Gaetani, Duke of Sermoneta, whose brilliant and witty conversation is unrivalled, and for whom I have had a very sincere friendship for many years. I found him lately as charming as ever, notwithstanding the cruel loss of his sight. The last time I ever dined out was at his house at Rome, when I was on my way to Naples in 1867.

* * * * *

John Gibson, the sculptor, the most guileless and amiable of men, was now a dear friend. His style was the purest Grecian, and had some of his works been found among the ruins, multitudes would have come to Rome to admire them. He was now in the height of his fame; yet he was so kind and encouraging to young people that he allowed my girls to go and draw in his studio, and one of my daughters, with a friend, modelled there for some time. His drawings for bas-reliefs were most beautiful. He drew very slowly, but a line once drawn was never changed. He ignored India-rubber or bread-crumbs, so perfect was his knowledge of anatomy, and so decided the character and expression he meant to give.

We had charades one evening in a small theatre in our house, which went off very well. There was much beauty at Rome at that time; no one who was there can have forgotten the beautiful and brilliant Sheridans. I recollect Lady Dufferin at the Easter ceremonies at St. Peter's, in her widow's cap, with a large black crape veil thrown over it, creating quite a sensation. With her exquisite features, oval face, and somewhat fantastical head-dress, anything more lovely could not be conceived; and the Roman people crowded round her in undisguised admiration of "la bella monaca Inglese." Her charm of manner and her brilliant conversation will never be forgotten by those who knew her. To my mind, Mrs. Norton was the most beautiful of the three sisters. Hers is a grand countenance, such as artists love to study. Gibson, whom I asked, after his return from England, which he had revisited after twenty-seven years' absence, what he thought of Englishwomen, replied, he had seen many handsome women, but no such sculptural beauty as Mrs. Norton's. I might add the Marchioness of Waterford, whose bust at Macdonald's I took at first for an ideal head, till I recognised the likeness.

Lady Davy used to live a great deal at Rome, and took an active part in society. She talked a great deal, and talked well when she spoke English, but like many of us had more pretension with regard to the things she could not do well than to those she really could. She was a Latin scholar, and as far as reading and knowing the literature of modern languages went she was very accomplished, but unfortunately, she fancied she spoke them perfectly, and was never happier than when she had people of different nations dining with her, each of whom she addressed in his own language. Many amusing mistakes of hers in speaking Italian were current in both Roman and English circles.

* * * * *

A few months were very pleasantly spent one summer at Perugia, where there is so much that is interesting to be seen. The neighbouring country is very beautiful, and the city being on the top of a hill is very cool during the hot weather. We had an apartment in the Casa Oddi-Baglioni—a name well known in Italian history—and I recollect spending some very pleasant days with the Conte Oddi-Baglioni, at a villa called Colle del Cardinale, some ten or twelve miles from the town. The house was large and handsomely decorated, with a profusion of the finest Chinese vases. On our toilet tables were placed perfumes, scented soap, and very elaborately embroidered nightdresses were laid out for use. I remember especially admiring the basins, jugs, &c., which were all of the finest japan enamel. There was a subterranean apartment where we dined, which was delightfully cool and pleasant, and at a large and profusely served dinner-table, while we and the guests with the owner of the house dined at the upper end, at the lower end and below the salt there were the superintendent of the Count's farms, a house decorator and others of that rank. It is not the only instance we met with of this very ancient custom. The first time Somerville and I came to Italy, years before this, while dining at a very noble house, the wet-nurse took her place, as a matter of course, at the foot of the dinner-table.

On the morning after our arrival and at a very

early hour there was a very fine eclipse of the sun, though not total at Perugia or the neighbourhood; the chill and unnatural gloom were very striking.

Perugia is one of the places in which the ancient athletic game of *pallone* is played with spirit. It is so graceful when well played that I wonder our active young men have not adopted it. A large leather ball filled with condensed air is struck and returned again by the opponent with the whole force of their right arms, covered to the elbow with a spiked wooden case. The promptness and activity required to keep up the ball is very great, and the impetus with which it strikes is such, that the boxes for spectators in the amphitheatres dedicated to this game are protected by strong netting. It is a very complicated game, and, I am told, somewhat resembles tennis.

* * * * *

On leaving Perugia we went for a few days to Asissi, spent a day at Chiusi, and then returned to Rome, which we found in a great state of excitement on account of three steamers which had just arrived from England to ply on the Tiber. The Pope and Cardinals made a solemn procession to bless them. No doubt they would have thought our method of dashing a bottle of wine on a vessel on naming her highly profane.

We constantly made expeditions to the country, to Tivoli, Veii, Ostia, &c., and my daughters rode on the Campagna. One day they rode to Albano, and on returning after dark they told me they had seen a most curious cloud which never altered its position; it was a very long narrow stripe reaching from the horizon till nearly over head—it was the tail of the magnificent comet of 1843.

We met with a great temptation in an invitation from Lady Stratford Canning, to go and visit them at Buyukdéré, near Constantinople, but *res arcta* prevented us from accepting what would have been so desirable in every respect. At this time I sat to our good friend Mr. Macdonald for my bust, which was much liked.*

* * * * *

One early summer we went to Loreto and Ancona, where we embarked for Trieste; the weather seemed fine when we set off, but a storm came on, with thunder and lightning, very high sea and several waterspouts. The vessel rolled and pitched, and we were carried far out of our course to the Dalmatian coast. I was obliged to remain a couple of days at Trieste to rest, and was very glad when we arrived

* The vessel on board which this bust was shipped for England ran on a shoal and sank, but as the accident happened in shallow water, the bust was recovered, none the worse for its immersion in salt water.

at Venice. The summer passed most delightfully at Venice, and we had ample time to see everything without hurry. I wrote very little this summer, for the scenery was so beautiful that I painted all day; my daughters drew in the Belle Arti, and Somerville had plenty of books to amuse him, besides sight-seeing, which occupied much of our time. In the Armenian convent we met with Joseph Warten, an excellent mathematician and astronomer; he was pastor at Neusatz, near Peterwardein in Hungary, and he was making a tour through Europe. He asked me to give him a copy of the "Mechanism of the Heavens," and afterwards wrote in Latin to Somerville and sent me some errors of the press he had met with in my book, but they were of no use, as I never published a second edition. We returned to Rome by Ravenna, where we stayed a couple of days, then travelled slowly along the Adriatic Coast. From thence we went by Gubbio and Perugia to Orvieto, one of the most interesting towns in Italy, and one seldom visited at that time; now the railway will bring it into the regular track of travellers.

* * * * *

[A few extracts from letters, written and received during this summer by my mother, may not be without interest. Also parts of two from my mother's old and valued friend Miss Joanna Baillie. The second letter was written several years later, and is nearly the last she ever wrote to my mother.

FROM MRS. SOMERVILLE TO WORONZOW GREIG, ESQ.

VENICE, 21st July, 1843.

I most sincerely rejoice to hear that Agnes and you have gone to the Rhine, as I am confident a little change of air and scene will be of the greatest service to you both. . . . . We are quite enchanted with Venice; no one can form an idea of its infinite loveliness who has not seen it in summer and in moonlight. I often doubt my senses, and almost fear it may be a dream. We are lodged to perfection, the weather has been charming, no oppressive heat, though the thermometer ranges from 75° to 80°, accompanied by a good deal of scirocco; there are neither flies nor fleas, and as yet the mosquitoes have not molested us. We owe much of our comfort to the house we are in, for there are scarcely any furnished lodgings, and the hotels are bad and dear, besides situation is everything at this season, when the smaller canals become offensive at low water, for, though there is little tide in the Mediterranean, there are four feet at new and full moon here, which is a great blessing. We have now seen everything, and have become acquainted with everybody, and met with kindness and attention beyond all description. Many of the great ducal families still exist, and live handsomely in their splendid palaces; indeed, the decay of Venice, so much talked of, is quite a mistake; certainly it is very different from what it was in its palmy days, but there is a good deal of activity and trade. The abolition of the law of primogeniture has injured the noble families more than anything else. We rise early, and are busy indoors all morning, except the girls, who go to the Academy of the *Belle Arti*, and paint from ten till three. We dine at four, and embark in our gondola at six or seven, and row about on the glassy sea till nine, when we go to the Piazza of San Marco, listen to a very fine military band, and sit gossiping till eleven or twelve, and then row home by the Grand Canal, or make a visit in one of the various houses that are open to us. One of the most remarkable of these is that of the Countess Mocenigo's, who has in one of her drawing-rooms the portraits of six doges of the Mocenigo name. I was presented by her to the Duc de Bordeaux, the other evening, a fat good-natured looking person. I was presented also to the Archduke—I forget what—son of the Archduke Charles, and admiral of the fleet here; a nice youth, but not clever. We meet him everywhere, and Somerville dined with him a few days ago. The only strangers of note are the Prince of Tour and Taxis, and Marshal Marmont. The Venetian ladies are very ladylike and agreeable, and speak beautifully. We have received uncommon kindness from Mr. Rawdon Brown; he has made us acquainted with everybody, as he is quite at home here, having been settled in Venice for several years, and has got a most beautiful house fitted up, in *rococo* style, with great taste; he is an adept at Venetian history. He supplies us with books, which are a great comfort. . . . . The other evening we were surprised by a perfect fleet of gondolas stopping under our windows, from one of which we had the most beautiful serenade; the moonlight was like day, and the effect was admirable. There was a *festa* the other night in a church on the water's edge; the shore was illuminated and hundreds of gondolas were darting along like swallows, the gondoliers rowing as if they had been mad, till the water was as much agitated as if there had been a gale of wind: nothing could be more animated. You will perceive from what I have said that the evening, till a late hour, is the time for amusement, in consequence of which I follow the Italian custom of sleeping after dinner, and am much the better for it. This place agrees particularly well with all of us, and is well suited for old people, who require air without fatigue. . . . .

Most affectionately,

MARY SOMERVILLE.

FROM MRS. SOMERVILLE TO WORONZOW GREIG, ESQ.

VENICE, *27th August*, 1843.

MY DEAR WORONZOW,

Your excellent letter, giving an account of your agreeable expedition up the Rhine, did not arrive till nearly a month after it was written. . . . . I regret exceedingly you could not stay longer, and still more that you could not come on and pay us a visit, and enjoy the charm of summer in Venice, so totally unlike every other place in every respect. I wished for you last night particularly. As we were leaving the Piazza San Marco, about eleven, a boat came up, burning blue lights, with a piano, violins, flutes, and about twenty men on board, who sang choruses in the most delightful manner, and sometimes solos. They were followed by an immense number of gondolas, and we joined the *cortége*, and all went under the Bridge of Sighs, where the effect was beautiful beyond description. We then all turned and entered the Grand Canal, which was entirely filled with

S 2

gondolas from one side to the other, jammed together, so that we moved *en masse*, and stopped every now and then to burn blue or red Bengal lights before the principal palaces, singing going on all the while. We saw numbers of our Venetian friends in their gondolas, enjoying the scene as much as we did, to whom it was almost new. I never saw people who enjoyed life more, and they have much the advantage of us in their delicious climate and aquatic amusements, so much more picturesque than what can be done on land. However, we have had no less than three dances lately. The Grand Duke of Modena, with his son and daughter-in-law, were here, and to them a *fête* was given by the Countess de Thurn. The palace was brilliant with lights; it is on the grand canal, and immediately under the balcony was a boat from which fireworks were let off, and then a couple of boats succeeded them, in which choruses were sung. The view from the balcony is one of the finest in Venice, and the night was charming, and there I was while the dancing went on. . . . . I never saw Somerville so well; this place suits us to the life, constant air and no fatigue; I never once have had a headache. . . . . Now, my dear W., tell me your tale; my tale is done.

Yours affectionately,

MARY SOMERVILLE.

FROM MRS. SOMERVILLE TO WORONZOW GREIG, ESQ.

ROME, PALAZZO LEPRI, VIA DEI CONDOTTI, 27*th October*, 1843.

MY DEAREST WORONZOW,

. . . . . We had a beautiful journey to Rome, with fine weather and no annoyance, notwithstanding the disturbed state of the country. At Padua we only remained long enough to see the churches, and it was impossible to pass within a few miles of Arquà without paying a visit to the house of Petrarch. At Ferrara we had a letter to the Cardinal Legate, who was very civil. His palace is the ancient abode of the house of Este. . . . . We had a long visit from him in the evening, and found him most agreeable; he regretted that there was no opera, as he would have been happy to offer us his box. Fourteen of those unfortunate men who have been making an attempt to raise an insurrection were arrested the day before; and the night before we slept at Lugo, the Carabineers had searched the inn during the night, entering the rooms where the people were sleeping. We should have been more than surprised to have been wakened by armed men at midnight. In travelling through Italy the *reliques* and history of the early Chris tians and of the Middle Ages have a greater attraction for me than those of either the Romans or Etruscans, interesting though these latter be, and in this journey my taste was amply gratified, especially at Ravenna, where the church of San Vitale and the Basilica of St. Apollinare in Classis, both built early in the 6th century, are the most magnificent specimens imaginable. Here also is the tomb of Theodore, a most wonderful building; the remains of his palace and numberless other objects of interest, too tedious to mention. Every church is full of them, and most valuable MSS. abound in the libraries. I like the history of the Middle Ages, because one feels that there is something in common between them and us; their names still exist in their descendants, who often inhabit the very palaces they dwelt in, and their very portraits, by the great masters, still hang in their halls; whereas we know nothing about the Greeks and Romans except their public deeds—their private life is a blank to us. Our journey through the Apennines was most beautiful, passing for days under the shade of magnificent oak forests or valleys rich in wine, oil, grain, and silk. We deviated from the main road for a short distance to Gubbio, to see the celebrated Eugubian tables, which are as sharp as if they had been engraved yesterday, but in a lost language. We stopped to rest at Perugia, but all our friends were at their country seats, which we regretted. The country round Perugia is unrivalled for richness and beauty, but it rained the morning we resumed our journey. It signified the less as we had been previously at Città della Pieve and Chiusi; so we proceeded to Orvieto in fine weather, still through oak forests. Orvieto is situated on the top of an escarped hill, very like the hill forts of India, and apparently as inaccessible; yet, by dint of numberless turns and windings, we did get up, but only in time for bed. Next morning we saw the sun rise on the most glorious cathedral. After all we had seen we were completely taken by surprise, and were filled with the highest admiration at the extreme beauty and fine taste of this remarkable building. . . . .

Your affectionate mother,

MARY SOMERVILLE.

FROM MISS JOANNA BAILLIE TO MRS. SOMERVILLE.

HAMPSTEAD, *December 27th*, 1843.

MY DEAR MRS. SOMERVILLE,

Besides being proud of receiving a letter from you, I was much pleased to know that I am, though at such a distance, sometimes in your thoughts. I was much pleased, too, with what you have said of the health and other gratifications you enjoy in Italy. I should gladly have thanked you at the time, had I known how to address my letter; and after receiving your proper direction from our friend Miss Montgomery, I have been prevented from using it by various things . . . . But though so long silent I have not been ungrateful, and thank you with all my heart. The account you give of Venice is very interesting. There is something affecting in still seeing the descendants of the former Doges holding a diminished state in their remaining palaces with so much courtesy. I am sure you have found yourself a guest in their saloons, hung with paintings of their ancestors, with very mixed feelings. However, Venice to the eye, as you describe it, is Venice still; and with its lights at night gleaming upon the waters makes a very vivid picture to my fancy. You no doubt have fixed it on canvas, and can carry it about with you for the delight of your friends who may never see the original.

In return to your kind inquiries after us, I have, all things considered, a very good account to give. Ladies of four score and upwards cannot expect to be robust, and need not be gay. We sit by the fire-side with our books (except when those plaguy notes are to be written) and receive the visits of our friendly neighbours very contentedly, and, I ought to say, and trust I may say, very thankfully. . . . . This morning brought one in whom I feel sure that you and your daughters take some interest, Maria Edgeworth. She has been dangerously ill, but is now nearly recovered, and is come from Ireland to pass the winter months with her sisters in London: weak in body, but the mind as clear and the spirits as buoyant as ever. You will be glad to hear that she even has it in her thoughts to write a new work, and has the plan of it nearly arranged. There will be nothing new in the story itself, but the purpose and treating of it will be new, which is, perhaps, a better thing. In our retired way of living, we know little of what goes on in the literary world. . . . . I was, however, in town for a few hours the other day, and called upon a lady of rank who has *fashionable* learned folks coming about her, and she informed me that there are new ideas regarding philosophy entertained in the world, and that Sir John Herschel was now considered as a slight, second-rate man, or person. Who are the first-rate she did not say, and, I suppose, you will not be much mortified to hear that your name was not mentioned at all. So much for our learning. My sister was much disappointed the other day when, in expectation of a ghost story from Mr. Dickens, she only got a grotesque moral allegory; now, as she delights in a ghost and hates an allegory, this was very provoking.

Believe me,

My dear Mrs. Somerville,

Yours with admiration and esteem,

J. BAILLIE.

FROM MISS JOANNA BAILLIE TO MRS. SOMERVILLE.

HAMPSTEAD, *January 9th*, 1851.

MY DEAR FRIEND,

My dear Mary Somerville, whom I am proud to call my friend, and that she so calls me. I could say much on this point, but I dare not. I received your letter from Mr. Greig last night, and thank you very gratefully. If my head were less confused I should do it better, but the pride I have in thinking of you as philosopher and a woman cannot be exceeded. I shall read your letter many times over. My sister and myself at so great an age are waiting to be called away in mercy by an Almighty Father, and we part with our earthly friends as those whom we shall meet again. My great monster book is now published, and your copy I shall send to your son who will peep into it, and then forward it to yourself. I beg to be kindly and respectfully remembered to your husband; I offer my best wishes to your daughters. . . . .

Yours, my dear Friend,
Very faithfully,
JOANNA BAILLIE.

My sister begs of you and all your family to accept her best wishes.

FROM SIR JOHN HERSCHEL TO MRS. SOMERVILLE.

18th March, 1844.

MY DEAR MRS. SOMERVILLE,

To have received a letter from you so long ago, and not yet to have thanked you for it, is what I could hardly have believed myself—if the rapid lapse of time in the uniform retirement in which we live were not pressed upon me in a variety of ways which convince me that as a man grows older, his sand, as the grains get low in the glass, slips through more glibly, and steals away with accelerated speed. I wish I could either send you a copy of my Cape observations, or tell you they are published or even in the press. Far from it—I do not expect to "go to press" before another year has elapsed, for though I have got my catalogues of Southern nebulæ and Double stars reduced and arranged, yet there is a great deal of other matter still to be worked through, and I have every description of reduction entirely to execute myself. These are very tedious, and I am a very slow computer, and have been continually taken off the subject by other matter, forced upon me by "pressure from without." What I am now engaged on is the monograph of the *principal* Southern Nebulæ, the object of which is to put on record every ascertainable particular of their actual appearance and the stars visible in them, so as to satisfy future observers whether *new stars* have appeared, or changes taken place in the nebulosity. To what an extent this work may go you may judge from the fact that the catalogue of visible stars actually mapped down in their places within the space of less than a square degree in the nebula about η Argus which I have just completed comprises between 1300 and 1400 stars. This is indeed a stupendous object. It is a vastly extensive branching and looped nebula, in the centre of the densest part of which is η Argus, itself a most remarkable star, seeing that from the fourth magnitude which it had in Ptolemy's time, it has risen (by *sudden starts,* and not gradually) to such a degree of brilliancy as *now* actually to surpass Canopus, and to be second only to Sirius. One of these *leaps* I myself witnessed when in the interval of ceasing to observe it in one year, and resuming its observation in two or three months after in the next, it had sprung over the heads of *all the stars of the first* magnitude, from Fomalhaut and Regulus (the two least of them) to α Centauri, which it then just equalled, and which is the brightest of all but Canopus and Sirius! It has since made a fresh jump—and who can say it will be the last?

One of the most beautiful objects in the southern hemisphere is a pretty large, perfectly round, and very well-defined planetary nebula, of a fine, full *independent* blue colour—the only object I have ever seen in the heavens fairly entitled to be called *independently* blue, *i.e.*, not by contrast. Another superb and most striking object is Lacaille's 30 Doradus, a nebula of great size in the larger nubicula, of which it is impossible to give a better idea than to compare it to a "true lover's knot," or assemblage of nearly circular nebulous loops uniting in a centre, in or near which is an exactly circular round dark hole. Neither this nor the nebula about η Argus have any, the slightest, resemblance to the representations given of them by Dunlop. . . . . As you are so kind as to offer to obtain information on any points interesting to me at Rome, here is one on which I earnestly desire to obtain the means of forming a correct opinion, *i.e.*, the *real* powers and merits of De Vico's great refractor at the Collegio Romano. De Vico's accounts of it appear to me to have not a little of the extra-marvellous in them. Saturn's *two* close satellites regularly observed—eight stars in the trapezium of Orion! α Aquilæ (as Schumacher inquiringly writes to me) divided into three! the supernumerary divisions of Saturn's ring well seen, &c., &c. And all by a Cauchoix refractor of eight inches? I fear me that these wonders are not for *female eyes*, the good monks are too well aware of the penetrating qualities of such optics to allow them entry within the sevenfold walls of their Collegio. Has Somerville ever looked through it? On his report I know I could quite rely. As for Lord Rosse's great reflector, I can only tell you what I hear, having never seen it, or even his three feet one. The great one is not yet completed. Of the other, those who *have* looked through it speak in raptures. I met not long since an officer who, at Halifax in Nova Scotia, saw *the comet* at noon close to the sun, and very conspicuous the day after the perihelion passage.

Your account of the pictures and other *deliciæ* of Venice makes our mouths water; but it is of no use, so we can only congratulate those who are in the full enjoyment of such things.

Ever yours most truly,
J. HERSCHEL.

On returning to Rome I was elected Associate of the College of Risurgenti, and in the following April I became an honorary member of the Imperial and Royal Academy of Science, Literature and Art at Arezzo. I finished an edition of the Physical Sciences, at which I had been working, and in spring Somerville hired a small house belonging to the Duca Sforza Cesarini, at Genzano, close to and with a beautiful view of the Lake of Nemi; but as I had not seen my son for some time, I now availed myself of the opportunity of travelling with our friend Sir Frederick Adam to England. We crossed the Channel at Ostend, and at the mouth of the Thames lay the old "Venerable," in which my father was flag-captain at the battle of Camperdown. I had a joyful meeting with my son and his wife, and we went to see many things that were new to me. One of our first expeditions was to the British Museum. I had already seen the Elgin marbles, and the antiquities collected at Babylon by Mr. Rich, when he was Consul at Bagdad, but now the Museum had been enriched by the marbles from Halicarnassus, and by the marvellous remains excavated by Mr. Layard from the ruins of Nineveh, the very site of which had been for ages unknown.

I frequently went to Turner's studio, and was always welcomed. No one could imagine that so much poetical feeling existed in so rough an exterior. The water-colour exhibitions were very good; my countrymen still maintained their superiority in that style of art, and the drawings of some English ladies were scarcely inferior to those of first-rate artists, especially those of my friend, Miss Blake, of Danesbury.

While in England I made several visits; the first was to my dear friends Sir John and Lady Herschel, at Collingwood, who received me with the warmest affection. I cannot express the pleasure it gave me to feel myself at home in a family where not only the highest branches of science were freely discussed, but where the accomplishments and graces of life were cultivated. I was highly gratified and proud of being godmother to Rosa, the daughter of Sir John and Lady Herschel. Among other places near Collingwood I was taken to see an excellent observatory formed by Mr. Dawes, a gentleman of independent fortune; and here I must remark, to the honour of my countrymen, that at the time I am writing, there are twenty-six private observatories in Great Britain and Ireland, furnished with first-rate instruments, with which some of the most important astronomical discoveries have been made.

[I received the following letter from my mother while we were at Genzano. It is one of several which record in her natural and unaffected words my mother's profound admiration for Sir John Herschel.

MRS. SOMERVILLE TO MISS SOMERVILLE.

SYDENHAM, *1st September*, 1844.
Sunday Night.

MY DEAR MARTHA,

. . . . We go to the Herschels' to-morrow, and there I shall finish this letter, as it is impossible to get it in time for Tuesday's post, but I have so much to do now that you must not expect a letter every post, and I had no time to begin this before, and I am too tired to sit up later to-night. . . . .

COLLINGWOOD, *Monday.*

This appears to be a remarkably beautiful place, with abundance of fine timber. . . . W. brought your dear nice letter; it makes me long to be with you, and, please God, I shall be so before long, as I set off this day fortnight.

Wednesday.

Yesterday I had a great deal of scientific talk with Sir John, and a long walk in the grounds which are extensive, and very pretty. Then the Airys arrived, and we had a large party at dinner. . . . . . I think, now, as I always have done, that Sir John is by much the highest and finest character I have ever met with; the most gentlemanly and polished mind, combined with the most exalted morality, and the utmost of human attainment. His view of everything is philosophic, and at the same time highly poetical, in short, he combines every quality that is admirable and excellent with the most charming modesty, and Lady Herschel is quite worthy of such a husband, which is the greatest praise I can give her. Their kindness and affection for me has been unbounded. Lady H. told me she heard such praises of you two that she is anxious to know you, and she hopes you will always look upon her and her family as friends. The christening went off as well as possible. Mr. Airy was godfather, and Mrs. Airy and I godmothers, but I had the naming of the child—Matilda Rose, after Lady Herschel's sister. I assure you I was quite adroit in taking the baby from the nurse and giving her to the clergyman. Sir John took Mrs. Airy and me a drive to see a very fine picturesque castle a few miles off. . . . I have got loads of things for experiments on light from Sir John with a variety of papers, and you may believe that I have profited not a little by his conversation, and have a thousand projects for study and writing, so I think painting will be at a standstill, only that I have promised to paint something for Lady Herschel. Sir John computes four or five hours every day, and yet his Cape observations will not be finished for two years. I have seen everything he is or has been doing.

Your affectionate mother,

MARY SOMERVILLE.

[My mother continues her recollections of this journey.

My next visit was to Lord and Lady Charles Percy, at Guy's Cliff, in Warwickshire, a pretty picturesque place of historical and romantic memory. The society was pleasant, and I was taken to Kenilworth and Warwick Castle, on the banks of the Avon, a noble place, still bearing marks of the Wars of the Roses. I never saw such magnificent oak-trees as those on the Leigh estate, near Guy's Cliff.

I then visited my maiden namesake, Mrs. Fairfax, of Gilling Castle, Yorkshire. She was a highly cultivated person, had been much abroad, and was a warm-hearted friend. I was much interested in the principal room, for a deep frieze surrounds the wall, on which are painted the coats of arms of all the families with whom the Fairfaxes have intermarried, ascending to very great antiquity; besides, every pane of glass in a very large bay window in the same room is stained with one of these coats of arms. Every morning after breakfast a prodigious flock of pea-fowl came from the woods around to be fed.

I now went to the vicinity of Kelso to visit my brother and sister-in-law, General and Mrs. Elliot, who lived on the banks of the Tweed. We went to Jedburgh, the place of my birth. After many years I still thought the valley of the Jed very beautiful; I fear the pretty stream has been invaded by manufactories: there is a perpetual war between civilization and the beauty of nature. I went to see the spot from whence I once took a sketch of Jedburgh Abbey and the manse in which I was born, which does not exist, I believe, now. When I was a very young girl I made a painting from this sketch. Our next excursion was to a lonely village called Yetholm, in the hills, some miles from Kelso, belonging to the gipsies. The "king" and the other men were absent, but the women were civil, and some of them very pretty. Our principal object in going there was to see a stone in the wall of a small and very ancient church at Linton, nearly in ruins, on which is carved in relief the wyvern and wheel, the crest of the Somervilles.

From Kelso I went to Edinburgh to spend a few days with Lord Jeffrey and his family. No one who had seen his gentle kindness in domestic life, and the warmth of his attachment to his friends, could have supposed he possessed that power of ridicule and severity which made him the terror of authors. His total ignorance of science may perhaps excuse him for having admitted into the "Review" Brougham's intemperate article on the undulatory theory of light, a discovery which has immortalized the name of Dr. Young. I found Edinburgh, the city of my early recollections, picturesque and beautiful as ever, but enormously increased both to the north and to the south. Queen Street, which in my youth was open to the north and commanded a view of the Forth and the mountains beyond, was now in the middle of the new town. All those I had formerly known were gone—a new generation had sprung up, living in all the luxury of modern times. On returning to London I spent a pleasant time with my son and his wife, who invited all those to meet me whom they thought I should like to see.

[My mother returned to Rome in autumn in company with an old friend and her daughter.

The winter passed without any marked event, but always agreeably; new people came, making a pleasant variety in the society, which, though still refined, was beginning to be very mixed, as was amusingly seen at Torlonia's balls and tableaux where many of the guests formed a singular contrast with the beautiful Princess, who was of the historical family of the Colonnas. I was often ashamed of my countrymen, who, all the while speaking of the Italians with contempt, tried to force themselves into their houses. Prince Borghese refused the same person an invitation to a ball five times. I was particularly scrupulous about invitations, and never asked for one in my life; nor did I ever seek to make acquaintances with the view of being invited to their houses.

* * * * *

[The following letters give a sketch of life during the summer months at Rome:—

MRS. SOMERVILLE TO W. GREIG, ESQ.

ROME, 3rd *August*, 1845.

MY DEAR WORONZOW,

. . . . I am glad you are so much pleased with my bust, and that it is so little injured after having been at the bottom of the sea. You will find Macdonald a very agreeable and original person. As to spending the summer in Rome, you may make yourself quite easy, for the heat is very bearable, the thermometer varying between 75° and 80° in our rooms during the day, which are kept in darkness, and at night it always becomes cooler. Thank God, we are all quite well, and Somerville particularly so; he goes out during the day to amuse himself, and the girls paint in the Borghese gallery. As for myself I have always plenty to do till half past three, when we dine, and after dinner I sleep for an hour or more, and when the sun is set we go out to wander a little, for a long walk is too fatiguing at this season. We have very little society, the only variety we have had was a very pretty supper party given by Signore Rossi, the French minister to the Prince and Princess de Broglie, son and daughter-in-law of the duke. The young lady is extremely beautiful, and as I knew the late Duchesse de Broglie (Madame de Staël's daughter) we soon got acquainted. They are newly married, and have come to spend part of the summer in Rome, so you see people are not so much alarmed as the English. . . We went yesterday evening to see the Piazza Navona full of water; it is flooded every Saturday and Sunday at this season; there is music, and the whole population of Rome is collected round it, carts and carriages splashing through it in all directions. I think it must be about three feet deep. It was there the ancient Romans had their naval games; and the custom of filling it with water in summer has lasted ever since. The fountain is one of the most beautiful in Rome, which is saying a great deal; indeed the immense gush of the purest water from innumerable fountains in every street and every villa is one of the peculiarities of Rome. I fear from what I have heard of those in Trafalgar Square that the quantity of water will be very miserable.

The papers (I mean the Times), are full of abuse of Mr. Sedgwick and Dr. Buckland, but their adversaries write such nonsense that it matters little. I do not think I have anything to add to my new edition. If you hear of anything of moment let me know. Perhaps something may have transpired at the British Association. . . . .

Your affectionate mother,

MARY SOMERVILLE.

MRS. SOMERVILLE TO W. GREIG, ESQ.

ROME, *May 28th*, 1845.

MY DEAR WORONZOW,

I don't know why I have so long delayed writing to you. I rather think it is because we have been living so quiet a life, one day so precisely similar to the preceding, that there has been nothing worth writing about. This is our first really summer-like day, and splendid it is; but we are sitting in a kind of twilight. The only means of keeping the rooms cool is by keeping the house dark and shutting out the external air, and then in the evening we have a delightful walk; the country is splendid, the Campagna one sheet of deep verdure and flowers of every kind in abundance. We generally have six or seven large nosegays in the room; we have only to go to some of the neighbouring villas and gather them. Most of the English are gone; people make a great mistake in not remaining during the hot weather, this is the time for enjoyment. We are busy all the morning, and in the afternoon we take our book or drawing materials and sit on the grass in some of the lovely villas for hours; then we come home to tea, and are glad to see anyone who will come in for an hour or two. We have had a son of Mr. Babbage here. He is employed in making the railway that is to go from Genoa to Milan, and he was travelling with eight other Englishmen who came to make arrangements for covering Italy with a network of these iron roads, connecting all the great cities and also the two seas from Venice to Milan and Genoa and from Ancona by Rome to Civita Vecchia. However the Pope is opposed to the latter part, but they say the cardinals and people wish it so much that he will at last consent . . . . Many thanks for the *Vestiges, &c.* I think it a powerful production, and was highly pleased with it, but I can easily see that it will offend in some quarters; however it should be remembered that there has been as much opposition to the true system of astronomy and to geological facts as there can be to this. At all events free and open discussion of all natural and moral phenomena must lead to truth at last. Is Babbage the author? I rather think he would not be so careful in concealing his name. . . .

[My mother made some curious experiments upon the effect of the solar spectrum on juices of plants and other substances, of which she sent an account to Sir John Herschel, who answered telling her that he had communicated her account of her experiments to the Royal Society.]

SIR JOHN HERSCHEL TO MRS. SOMERVILLE.

COLLINGWOOD, *November* 21*st*, 1845.

MY DEAR MRS. SOMERVILLE,

I cannot express to you the pleasure I experienced from the receipt of your letter and the perusal of the elegant experiments it relates, which appear to me of the highest interest and show (what I always suspected), that there is a world of wonders awaiting disclosure in the solar spectrum, and that influences widely differing from either light, heat or colour are transmitted to us from our central luminary, which are mainly instrumental in evolving and maturing the splendid hues of the vegetable creation and elaborating the juices to which they owe their beauty and their vitality. I think it certain that heat goes for something in evaporating your liquids and thereby causing some of your phenomena; but there is a difference of *quality* as well as of *quantity* of heat brought into view which renders it susceptible of analysis by the coloured juices so that in certain parts of the spectrum it is retained and fixed, in others reflected according as the nature of the tint favours the one or the other. Pray go on with these delightful experiments. I wish you could save yourself the fatigue of watching and directing your sunbeam by a clock work. If I were at your elbow I could rig you out a heliotrope quite sufficient with the aid of any common wooden clock. . . . . . . Now I am going to take a liberty (but not till after duly consulting Mr. Greig with whose approbation I act, and you are not to gainsay our proceedings) and that is to communicate your results in the form of "an extract of a letter" to myself—to the Royal Society. You may be very sure that I would not do this if I thought that the experiments were not intrinsically quite deserving to be recorded in the pages of the Phil. Trans. and if I were not sure that they will lead to a vast field of curious and beautiful research; and as you have already once contributed to the Society, (on a subject connected with the spectrum and the sunbeam) this will, I trust, not appear in your eyes in a formidable or a repulsive light, and it will be a great matter of congratulation to us all to know that these subjects continue to engage your attention, and that you can turn your residence in that sunny clime to such admirable account. So do not call upon me to retract (for before you get this the papers will be in the secretary's hands).

I am here nearly as much out of the full stream of scientific matters as you at Rome. We had a full and very satisfactory meeting at Cambridge of the British Association, with a full attendance of continental magnetists and meteorologists, and within these few days I have learned that our Government meant to grant all our requests and continue the magnetic and meteorological observations. Humboldt has sent me his Cosmos (Vol. I.), which is good, all but the first 60 pages, which are occupied in telling his readers what his book is *not* to be. Dr. Whewell has just published *another* book on the Principles of Morals, and also *another* on education, in which he cries up the geometrical processes in preference to analysis. . . . . .

Yours very faithfully,
J. HERSCHEL.

The Prince and Princesse de Broglie came to Rome in 1845, and Signore Pellegrino Rossi, at this time French Minister at the Vatican, gave them a supper party, to which we were invited. We had met with him long before at Geneva, where he had taken refuge after the insurrection of 1821. He was greatly esteemed there and admired for his eloquence in the lectures he gave in the University. It was a curious circumstance, that he, who was a Roman subject, and was exiled, and, if I am not mistaken, condemned to death, should return to Rome as French Minister. He had a remarkably fine countenance, resembling some ancient Roman bust. M. Thiers had brought in a law in the French Chambers to check the audacity of the Jesuits, and Rossi was sent to negotiate with the Pope. We had seen much of him at Rome, and were horrified, in 1848, to hear that he had been assassinated on the steps of the Cancelleria, at Rome, where the Legislative Assembly met, and whither he was proceeding to attend its first meeting. No one offered to assist him, nor to arrest the murderers except Dr. Pantaleone, a much esteemed Roman physician, and member of the Chamber, who did what he could to save him, but in vain; he was a great loss to the Liberal cause.

Towards the end of summer we spent a month most agreeably at Subiaco, receiving much civility from the Benedictine monks of the Sacro Speco, and visiting all the neighbouring towns, each one perched on some hill-top, and one more romantically picturesque than the other. It was in this part of the country that Claude Lorraine and Poussin studied and painted. I never saw more beautiful country, or one which afforded so many exquisite subjects for a landscape painter. We went all over the country on mules—to some of the towns, such as Cervara, up steep flights of steps cut in the rock. The people, too, were extremely picturesque, and the women still wore their costumes, which probably now they have laid aside for tweeds and Manchester cottons.

I often during my winters in Rome went to paint from nature in the Campagna, either with Somerville or with Lady Susan Percy, who drew very prettily. Once we set out a little later than usual, when, driving through the Piazza of the Bocca della Verità, we both called out, "Did you see that? How horrible!" It was the guillotine; an execution had just taken place, and had we been a quarter of an hour earlier we should have passed at the fatal moment. Under Gregory XVI. everything was conducted in the most profound secrecy; arrests were made almost at our very door, of which we knew nothing; Mazzini was busily at work on one side, the Jesuitical party actively intriguing, according to their wont, on the other; and in the mean time society went on gaily at the surface, ignorant of and indifferent to the course of events. We were preparing to leave Rome when Gregory died. We put off our journey to see his funeral, and the Conclave, which terminated, in the course of scarcely two days, in the election of Pius IX.

We also saw the new Pope's coronation, and witnessed the beginning of that popularity which lasted so short a time. Much was expected from him, and in the beginning of his reign the moderate liberals fondly hoped that Italy would unite in one great federation, with Pius IX. at the head of it; entirely forgetting how incompatible a theocracy or government by priests ever must be with all progress and with liberal institutions. Their hopes were soon blighted, and after all the well-known events of 1848 and 1849, a reaction set in all over Italy, except in gallant little Piedmont, where the constitution was maintained, thanks to Victor Emmanuel, and especially to that great genius, Camillo Cavour, and in spite of the disastrous reverses at Novara. Once more in 1859 Piedmont went to war with Austria, this time with success, and with the not disinterested help of France. One province after another joined her, and Italy, freed from all the little petty princes, and last, not least, from the Bourbons, has become that one great kingdom which was the dream of some of her greatest men in times of old.

We went to Bologna for a short time, and there the enthusiasm for the new Pope was absolutely intolerable. "Viva Pio Nono!" was shouted night and day. There was no repose; bands of music went about the streets, playing airs composed for the occasion, and in the theatres it was even worse, for the acting was interrupted, and the orchestra called upon to play the national tunes in vogue, and repeat them again and again, amid the deafening shouts and applause of the excited audience. We found the Bolognese very sociable, and it was by far the most musical society I ever was in. Rossini was living in Bologna, and received in the evening, and there was always music, amateur and professional, at his house. Frequently there was part-singing or choruses, and after the music was over the evening ended with a dance. We frequently saw Rossini some years later, when we resided at Florence. He was clever and amusing in conversation, but satirical. He was very bitter against the modern style of opera-singing, and considered the singers of the present day, with some exceptions, as wanting in study and finish. He objected to much of the modern music, as dwelling too constantly on the highest notes of the voice, whereby it is very soon deteriorated, and the singer forced to scream ; besides which, he considered the orchestral accompaniments too loud. I, who recollected Pasta, Malibran, Grisi, Rubini, and others of that epoch, could not help agreeing with him when I compared them to the singers I heard at the

Pergola and elsewhere. The theatre, too, was good at Bologna, and we frequently went to it.

One evening we were sitting on the balcony of the hotel, when we saw a man stab another in the back of the neck, and then run away. The victim staggered along for a minute, and then fell down in a pool of blood. He had been a spy of the police under Gregory XVI., and one of the principal agents of his cruel government. He was so obnoxious to the people that his assassin has never been discovered.

From Bologna we went for a few weeks to Recoaro, where I drank the waters, after which we travelled to England by the St. Gothard pass.

CHAPTER XVI.

PUBLISHES "PHYSICAL GEOGRAPHY"—LETTER FROM HUMBOLDT—CHRISTMAS AT COLLINGWOOD—LETTER FROM MRS. SOMERVILLE—FARADAY—LETTER FROM FARADAY—KEITH JOHNSTONE'S MAPS—WINTER AT MUNICH—SALZBURG—LAKE OF GARDA—MINISCALCHI—POEM BY CATERINA BRENZONI—LETTER FROM BRENZONI—LETTER FROM MRS. SOMERVILLE—ÉLOGE BY MINISCALCHI—WINTER AT TURIN—BARON PLANA—CAMILLO CAVOUR—COLLINE NEAR TURIN—GENOA—TERESA DOVIA—FLORENCE—MISS F. P. COBBE—VIVISECTION—EXCURSIONS IN THE NEIGHBOURHOOD—CHOLERA—MISERICORDIA—PIO NONO IN TUSCANY—COMET—TUSCAN REVOLUTION—WAR IN LOMBARDY—ENTRY OF VICTOR EMMANUEL INTO FLORENCE—LETTERS FROM MRS. SOMERVILLE—MY FATHER'S DEATH—LETTER FROM MISS COBBE.

WE spent the autumn in visiting my relations on the banks of the Tweed. I was much out of health at the time. As winter came on I got better, and was preparing to print my "Physical Geography" when "Cosmos" appeared. I at once determined to put my manuscript in the fire when Somerville said, "Do not be rash—consult some of our friends—Herschel for instance." So I sent the MS. to Sir John Herschel, who advised me by all means to publish it. It was very favourably reviewed by Sir Henry Holland in the "Quarterly," which tended much to its success. I afterwards sent a copy of a later edition to Baron Humboldt, who wrote me a very kind letter in return.

BARON HUMBOLDT TO MRS. SOMERVILLE

A SANS SOUCI, *ce* 12 *Juillet*, 1849.

MADAME,

C'est un devoir bien doux à remplir, Madame, que de vous offrir l'hommage renouvellé de mon dévouement et de ma respectueuse admiration. Ces sentimens datent de bien loin chez l'homme antidiluvien auquel vous avez daigné adresser des lignes si aimables et la nouvelle édition de ce bel ouvrage qui m'a charmé et *instruit* dès qu'il avait paru pour la première fois. A cette grande supériorité que vous possedez et qui a si noblement illustré votre nom, dans les hautes régions de l'analyse mathématique, vous joignez, Madame, une variété de connaissances dans toutes les parties de la physique et de l'histoire naturelle descriptive. Après votre "Mechanism of the Heavens," le philosophique ouvrage "Connexion of the Physical Sciences" avait été l'objet de ma constante admiration. Je l'ai lu en entier et puis relu dans la septième édition qui a paru en 1846 dans les tems où nous étions plus calme, où l'orage politique ne grondait que de loin. L'auteur de l'imprudent "Cosmos" devoit saluer plus que tout autre la "Géographie Physique" de Mary Somerville. J'ai su me la procurer dès les premières semaines par les soins de notre ami commun le Chev. Bunsen. Je ne connais dans aucune langue un ouvrage de Géographie physique

que l'on pourrait comparer au votre. Je l'ai de nouveau étudié dans la dernière édition que je dois à votre gracieuse bienveillance. Le sentiment de précision que vos habitudes de "Géomètre" vous ont si profondement imprimé, pénètre tous vos travaux, Madame. Aucun fait, aucune des grandes vues de la nature vous échappent. Vous avez profité et des livres et des conversations des voyageurs dans cette malheureuse Italie où passe la grande route de l'Orient et de l'Inde. J'ai été surpris de la justice de vos aperçus sur la Géographie des plantes et des animaux. Vous dominez dans ces régions comme en astronomie, en météorologie, en magnetisme. Que n'ajoutez-vous pas la sphère céleste, l'uranològie, votre patrimoine, à la sphère terrestre? C'est vous seule qui pourriez donner à votre belle littérature un ouvrage cosmologique original, un ouvrage écrit avec cette lucidité et ce goût que distingue tout ce qui est emané de votre plume. On a, je le sais, beaucoup de bienveillance pour mon Cosmos dans votre patrie; mais il en est des *formes* de composition littéraires, comme de la variété des races et de la différence primitive des langues. Un ouvrage traduit manque de vie; ce que plait sur les bords du Rhin doit paraître bizarre sur les bords de la Tamise et de la Seine. Mon ouvrage est une production essentiellement allemande, et ce caractère même, j'en suis sûr, loin de m'en plaindre lui donne le goût du terroir. Je jouis d'une bonne fortune à laquelle (à cause de mon long séjour en France, de mes prédilections personnelles, de mes hérésies politiques) le *Léopard* ne m'avait pas trop accoutumé. Je demande à l'illustre auteur du volume sur la Mécanique Céleste d'avoir le courage d'aggrandir sa Géographie Physique. Je suis sûr que le grand homme que nous aimons le plus, vous et moi, Sir

John Herschel, serait de mon opinion. Le MONDE, je me sers du titre que Descartes voulait donner à un livre dont nous n'avons que de pauvres fragmens; le *Monde* doit être écrit pour les Anglais par un auteur de race pure. Il n'y a pas de sève, pas de vitalité dans les traductions les mieux faites. Ma santé s'est conservé miraculeusement à l'âge de quatre-vingts ans, de mon ardeur pour le travail nocturne au milieu des agitations d'une position que je n'ai pas besoin de vous depeindre puisque l'excellente Mademoiselle de —— vous l'a fait connaître. J'ai bouleversé, changé mes deux volumes des "Ansichten." Il n'en est resté que ¼. C'est comme un nouvel ouvrage que j'aurai bientôt le bonheur de vous adresser si M. Cotta pense pouvoir hasarder une publication dans ces tems où la force physique croit guérir un mal moral et *vacciner* le contentement à l'Allemagne unitaire!! Le troisième volume de mon Cosmos avance, mais la sérénité manque aux âmes moins crédules.

Agréez, je vous supplie, l'hommage de mon affectueuse et respectueuse reconnaissance,

ALEXANDRE DE HUMBOLDT.

Somerville and I spent the Christmas at Collingwood with our friends the Herschels. The party consisted of Mr. Airy, Astronomer-Royal, and Mr. Adams, who had taken high honours at Cambridge. This young man and M. Leverrier, the celebrated French astronomer, had separately calculated the orbit of Neptune and announced it so nearly at the same time, that each country claims the honour of the discovery. Mr. Adams told Somerville that the following sentence in the sixth edition of the "Connexion of the Physical Sciences," published in the year 1842, put it into his head to calculate the orbit of Neptune. "If after the lapse of years the tables formed from a combination of numerous observations should be still inadequate to represent the motions of Uranus, the discrepancies may reveal the existence, nay, even the mass and orbit of a body placed for ever beyond the sphere of vision." That prediction was fulfilled in 1846, by the discovery of Neptune revolving at the distance of 3,000,000,000 of miles from the sun. The mass of Neptune, the size and position of his orbit in space, and his periodic time, were determined from his disturbing action on Uranus before the planet itself had been seen.

We left Collingwood as ever with regret.

[The following is an extract from a letter written by my mother during this visit:—

FROM MRS. SOMERVILLE TO W. GREIG, ESQ.

COLLINGWOOD, 1*st January*, 1848.

. . . . You can more easily conceive than I can describe the great kindness and affection which we have received from both Sir John and Lady Herschel; I feel a pride and pleasure beyond what I can express in having such friends. Collingwood is a house by itself in the world, there certainly is nothing like it for all that is great and good. The charm of the conversation is only equalled by its variety—every subject Sir John touches turns to doubly refined gold; profound, brilliant, amiable, and highly poetical, I could never end admiring and praising him. Then the children are so nice and he so kind and amusing to them, making them quite his friends and companions.

Yours, my dearest Woronzow,
Most affectionately,
M. SOMERVILLE.

We had formed such a friendship with Mr. Faraday that while we lived abroad he sent me a copy of everything he published, and on returning to England we renewed our friendship with that illustrious philosopher, and attended his lectures at the Royal Institution. He had already magnetized a ray of polarised light, but was still lecturing on the magnetic and diamagnetic properties of matter. At the last lecture we attended he showed the diamagnetism of flame, which had been proved by a foreign philosopher. Mr. Faraday never would accept of any honour; he lived in a circle of friends to whom he was deeply attached. A touching and beautiful memoir was published of him by his friend and successor, Professor Tyndall, an experimental philosopher of the very highest genius.

[The following letter was the last my mother received from Faraday:—

FROM PROFESSOR FARADAY TO MRS. SOMERVILLE.

ROYAL INSTITUTION, *17th January*, 1859.

MY DEAR MRS. SOMERVILLE,

So you have remembered me again, and I have the delight of receiving from you a new copy of that work which has so often instructed me; and I may well say, cheered me in my simple homely course through life in this house. It was most kind to think of me; but ah! how sweet it is to believe that I have your *approval* in matters where kindness would be nothing, where judgment alone must rule. I almost doubt myself when I think I have your approbation, to some degree at least, in what I may have thought or said about gravitation, the forces of nature, their conservation, &c. As it is, I *cannot* go back from these thoughts; on the contrary, I feel encouraged to go on by way of experiment, but am not so able as I was formerly; for when I try to hold the necessary group of thoughts in mind at one time, with the judgment suspended on almost all of them, then my head becomes giddy, and I am obliged to lay all aside for a while. I am trying for *time* in magnetic action, and do not despair of reaching it, even though it may be only that of light. *Nous verrons.*

I have been putting into one volume various papers

of mine on experimental branches in chemistry and physics. The index and title-page has gone to the printer, and I expect soon to receive copies from him. I shall ask Mr. Murray to help me in sending one to you which I hope you will honour by acceptance. There is nothing new in it, except a few additional pages about "*regelation*," and also "gravity." It is useful to get one's scattered papers together with an index, and society seems to like the collection sufficiently to pay the expenses. . . . . Pray remember me most kindly to all with whom I may take that privilege, and believe me to be most truly,

Your admirer and
faithful servant,
M. FARADAY.

My mother wrote of this letter:—

FLORENCE, 8*th February*, 1859.

. . . . I have had the most charming and gratifying letter from Faraday; I cannot tell you how I value such a mark of approbation and friendship from the greatest experimental philosopher and discoverer next to Newton.

We returned to the continent in autumn, so I could not superintend the publication of my "Physical Geography," but Mr. Pentland kindly undertook to carry it through the press. Though I never was personally acquainted with Mr. Keith Johnston, of

Edinburgh, that eminent geographer gave me copies of both the first and second editions of his splendid "Atlas of Physical Geography," which were of the greatest use to me. Besides, he published some time afterwards a small "School Atlas of Ancient, Modern, and Physical Geography," intended to accompany my work; obligations which I gratefully acknowledge. No one has attempted to copy my "Connexion of the Physical Sciences," the subjects are too difficult; but soon after the publication of the "Physical Geography" a number of cheap books appeared, just keeping within the letter of the law, on which account it has only gone through five editions. However a sixth is now required.

* * * * *

The moment was unfavourable for going into Italy, as war was raging between Charles Albert and the Austrians, so we resolved to remain at Munich, and wait the course of events. We got a very pretty little apartment, well furnished with stoves, and opposite the house of the Marchese Fabio Pallavicini, formerly Sardinian minister at Munich. We spent most of our evenings very pleasantly at their house. We attended the concerts at the Odeon of classical music: the execution was perfect, but the music was so refined and profound that it passed my comprehension, and I thought it tedious. The hours at

Munich were so early that the opera ended almost at the time it began in London.

In the spring we went to Salzburg, where we remained all summer. We had an apartment in a dilapidated old château, about an hour's walk from the town, called Leopold's Krone. The picturesque situation of the town reminded me of the Castle and Old Town of Edinburgh. The view from our windows was alpine, and the trees bordering the roads were such as I have rarely seen out of England. We made many excursions to Berchtesgaden, where King Louis and his court were then living, and went to the upper end of the Königsee. I have repeatedly been at sea in very stormy weather without the smallest idea of fear; but the black, deep water of this lake, under the shadow of the precipitous mountains, made a disagreeable impression on me. I thought if I were to be drowned I should prefer the blue sea to that cold, black pool. The flora was lovely, and on returning from our expeditions in the evening, the damp, mossy banks were luminous with glowworms: I never saw so many, either before or since. We never fail to make acquaintances wherever we go, and our friends at Munich had given us letters to various people who were passing the summer there, many of whom had evening receptions once a week. At the

Countess Irene Arco's beautiful Gothic château of Anif, which rises out of a small pellucid lake, and is reached by a bridge, we spent many pleasant evenings, as well as at Countess Bellegarde's, and at Aigen, which belonged to the Cardinal Schwartzenberg. We never saw him, but went to visit his niece, with whom we were intimate.

The war being over, we went by Innsbruck and the Brenner to Colà, on the Lago di Garda, within five miles of Peschiera, where we spent a month with Count and Countess Erizzo Miniscalchi, who had been our intimate friends for many years. The devastation of the country was frightful. Peschiera and its fortifications were in ruins; the villages around had been burnt down, and the wretched inhabitants were beginning to repair their roofless houses. Our friends themselves had but recently returned to Colà, which, from its commanding situation, was always the head-quarters of whatever army was in possession of the country around. On this account, the family had to fly more than once at the approach of the enemy. In 1848 the Countess had fled to Milan, and was confined at the very time the Austrians under Radetsky were besieging the town, which was defended by Charles Albert. Fearing what might occur when the city was surrendered, the lady, together with her new-born infant and the rest of her family, escaped the next day with considerable difficulty, and travelled to Genoa.

Although not acquainted with quite so many languages as Mezzofanti, Count Miniscalchi is a remarkable linguist, especially with regard to Arabic and other oriental tongues. He has availed himself of his talent, and published several works, the most interesting of which is a translation of the Gospel of St. John from Syro-Chaldaic (the language probably spoken by our Saviour) into Latin. The manuscript, from which this translation is made, is preserved in the Vatican.

[While we were at Colà my mother received a visit from a very distinguished and gifted lady, the Countess Bon-Brenzoni. As an instance of the feelings entertained by an Italian woman towards my mother, I insert a letter written by the Countess some time afterwards, and also an extract from her poems:—

FROM THE COUNTESS BON-BRENZONI TO MRS. SOMERVILLE.

VERONA, 28 *Maggio*, 1853.

ILLUSTRE SIGNORA,

Fui molto contenta udendo che finalmente le sia giunto l' involto contenente le copie stampate del Carme, ch' ebbi l' onore di poterle offerire, mentre io era in gran pensiero non forse fossero insorte difficoltà, o ritardi, in causa della posta. Ma, ben più che per questo la sua graziosissima lettera mi fù di vera consolazione, per l' accoglienza tutta benevola e generosa ch' Ella fece a' miei versi. La ringrazio delle parole piene di bontà ch' Ella mi scrive, e di aversi preso la gentil cura di farlo in italiano; così potess' io ricambiarla scrivendo a Lei in inglese! Pur mi conforta la certezza che il linguaggio delle anime sia uno solo; mentre io non so s' io debba chiamar presunzione, o ispirazione questa, che mi fa credere, che esista fra la sua e la mia una qualche intelligenza, e quantunque i suoi meriti e la sua bontà me ne spieghino in gran parte il mistero, pure trovo essere cosa non comune questo pensiero, che al mio cuore parla di Lei incessantemente, da quel giorno ch' io l' ho veduta per la prima e l' unica volta!

Ah se è vero che fra i sentimenti di compiacenza ch' Ella provò per gli elogi ottenuti de' suoi lavori, abbia saputo trovar luogo fra i più cari quello che le destò nell' animo l' espressione viva e sincera della mia ammirazione e del mio umile affetto, io raggiunsi un punto a cui certo non avea osato aspirare!

Il trovarmi con Lei a Colà, od altrove che fosse, è uno de' miei più cari desideri, e son lieta delle sue parole che me ne danno qualche speranza.

Voglia presentare i miei distinti doveri all' eccelente suo Sig^{re} marito ed alle amabili figlie; e mentre io le prego da Dio le più desiderabili benedizioni, Ella si ricordi di me siccome di una persona, che sebbene lontana fisicamente, le è sempre vicina coll' animo, nei sentimenti della più affetuosa venerazione.

Incoraggiata dalla sua bontà, mi onoro segnarmi

amica affezionatissima

CATERINA BON-BRENZONI.

The "Carme" spoken of in the above letter form a long poem on modern astronomy, entitled "I Cieli," (published by Vallardi. Milan: 1853). The opening lines contain the following address to Mrs. Somerville,—doubtless a genuine description of the author's feelings on first meeting the simple-mannered lady whose intellectual greatness she had long learned to appreciate:—

Donna, quel giorno ch' io ti vidi in prima,
Dimmi, hai Tu scôrto sul mio volto i segni
Dell' anima commossa ?—Hai Tu veduto
Come trepida innanzi io ti venia,
E come reverenza e maraviglia
Tenean sospesa sull' indocil labbro
La parola mal certa ?—Ah! dimmi, hai scôrto
Come fur vinte dall' affetto allora
Che t'udii favellar soave e piana,
Coll' angelica voce e l' umiltade,
Che a' suoi più cari sapïenza insegna ?—
Questa, io dicea tra me, questa è Colei,
Di che le mille volte udito ho il nome
Venerato suonar tra i più famosi ?
Questa è Colei che negli eterei spazj
Segue il cammin degli astri, e ne misura
Peso, moto, distanza, orbita e luce ?

* * * *

Another record of our visit to Colà is in a letter of my mother to my brother.

MRS. SOMERVILLE TO W. GREIG, ESQ.

TURIN, *4th Dec.*, 1849.

MY DEAREST WORONZOW,

We arrived here all well the day before yesterday, after a fair but bitterly cold journey, bright sunshine and keen frost, and to-day we have a fall of snow. . . . . . It was a great disappointment not finding letters here, and I fear many have been lost on both sides, though we took care not to touch on political events, as all letters are opened by the Austrian police in Lombardy. We spent five weeks with our friends the Miniscalchis very agreeably, and received every mark of kindness and hospitality. They only live at Verona during the winter, and we found them in their country house at Colà situated on a height overlooking the Lago di Garda, with the snowy Alps on the opposite side of the lake. The view from their grounds is so fine that I was tempted to paint once more. They took us to see all the places in the neighbourhood; often a sad sight, from having been the seat of war and siege. The villages are burnt and the churches in ruin. But the people are repairing the mischief as fast as possible, and the fields are already well cultivated. The Count is a man of great learning and is occupied in the comparison of languages, especially the Eastern; he knows twenty-four and speaks Arabic as fluently as Italian. He is in the habit of speaking both Arabic and Chaldee every day, as there is a most learned Chaldean priest living with them, whose conversation gave me great pleasure and much information. The Count has moreover a black servant who speaks these languages, having been bought by the Count during his long residence in the East, and is now treated like one of the family. I obtained much information which will be useful in my next edition of the Physical Geography . . . . . .

Your affectionate mother,

MARY SOMERVILLE.

[After my mother's death, our old friend Count Miniscalchi made a beautiful and touching "éloge" on her at a meeting of the Royal Italian Geographical Society, to a numerous audience assembled in the great hall of the Collegio Romano at Rome.

My mother was an honorary member of this Society, besides which the first gold medal granted by them was voted by acclamation to her. Her Recollections continue as follows:—

From Colà we went to Turin, where I became personally acquainted with Baron Plana, Director of the Observatory. He had married a niece of the illustrious mathematician La Grange, who proved the stability of the solar system. Plana, himself, was a very great analyst; his volume on the Lunar Perturbations is a work of enormous labour. He gave me a copy of it and of all his works; for I continued to have friendly intercourse with him as long as he lived. As soon as he heard of our arrival, he came to take us out to drive. I never shall forget the beauty of the Alps, and the broad valley of the Po and Dora, deeply covered with snow, and sparkling in bright sunshine. Another day the Baron took us to a church, from the cupola of which a very long pendulum was swinging, that we might see the rotation of the earth visibly proved by its action on the pendulum, according to M. Foucault's experiment. He devoted his time to get us established, and we found a handsome apartment in Casa Cavour, and became acquainted with both the brothers to whom it belonged. Count Camillo Cavour, then Minister of the Interior, was the only great statesman Italy ever produced in modern times. His premature death is deplorably felt at the present day. He was a real genius, and the most masterly act of his administration was that of sending an army to act in concert with the French and English in the Crimean war. By it he at once gave Italy the rank of an independent European power, which was the first step towards Italian unity. He was delightful and cheerful in society, and extremely beloved by his family and friends.

* * * * *

In spring we hired a villa on the Colline above Turin. The house was in a garden, with a terrace, whence the ground sank rapidly to the plain; low hills, clothed with chestnut forests, abounding in lilies of the valley, surrounded us behind. The summer had been stormy, and one evening we walked on the terrace to look at the lightning, which was very fine, illuminating the chain of Alps. By-and-by it ceased, and the darkness was intense; but we continued to walk, when, to our surprise, a pale bluish light rose in the Val di Susa, which gradually spread along the summit of the Alps, and the tops of the hills behind our house; then a column of the same pale blue light, actually within our reach, came curling up from the slope close to the terrace, exactly as if wet weeds had been burning. In about ten minutes the whole vanished; but in less than a quarter of an hour the phenomena were repeated exactly as described, and were followed by a dark night and torrents of rain. It was a very unusual instance of what is known as electric glow; that is, electricity without tension.

On our road to Genoa, we went to see some kind Piedmontese friends, who have a château in the Monferrat, not many miles from Asti, where we left the railroad. We had not gone many miles when the carriage we had hired was upset, and, although nobody had broken bones, I got so severe a blow on my forehead that I was confined to bed for nearly a month, and my face was black and blue for a much longer time. Nothing could equal the unwearied kindness of our friends during my illness.

When I was able to travel, we went to Genoa for the winter, and lived on the second floor of a large house on the Acqua Sola, and overlooking the sea. Here first began our friendship with the Marchesa Teresa Doria, whose maiden name was Durazzo; in her youth one of the handsomest women in Genoa, a lady distinguished for her generous character and cultivated mind, and who fearlessly avowed her opinions at a time when it was a kind of disgrace to be called a Liberal. Her youngest son, Giacomo, has devoted his life to the study of natural history, and his mother used all her influenee to encourage and help him in a pursuit so unusual amongst people of rank in this country. Later, he travelled in Persia for two years, to make collections, and since then resided for a long time in Borneo, and is now arranging a museum in his native city. The Marchesa has always been a warm and devoted friend to me and mine.

It was here that we got our dear old parrot Lory, who is still alive and merry.

* * * * *

Our next move was to Florence, where we already knew many people. We had a lease of a house in Via del Mandorlo, which had a small garden and a balcony, where we often sat and received in the warm summer evenings. My daughters had adorned it and the garden with rare creepers, shrubs, and flowers.

We had a visit from our friend Gibson, as he passed through Florence on his way to Switzerland. He told us the history of his early life, as given in his biography; and much that is not mentioned there. He was devotedly attached to the Queen, and spoke of her in his simple manner as a charming lady.

Miss Hosmer was travelling with Gibson, an American young lady, who was his pupil, and of whose works he was very proud. He looked upon her as if she had been his daughter, and she took care of him; for he was careless and forgetful when travelling. I have the sincerest pleasure in expressing my admiration for Miss Hosmer, who has proved by her works that our sex possesses both genius and originality in the highest branches of art.

It was at Florence that I first met my dear friend and constant correspondent, Frances Power Cobbe. She is the cleverest and most agreeable woman I ever met with, and one of the best. There is a distant connection between us, as one of her ancestors married a niece of Lord Fairfax, the Parliamentary general, many of whose letters are in the possession of her family. A German professor of physiology at Florence roused public indignation by his barbarous vivisections, and there was a canvass for a Memorial against this cruel practice. Miss Cobbe took a leading part in this movement, and I heartily joined, and wrote to all my acquaintances, requesting their votes; among others, to a certain Marchese, who had published something on agriculture. He refused his vote, saying, "Perhaps I was not aware that the present state of science was one of induction." Then he went on explaining to me what "induction" meant, &c., &c., which amused me not a little. It made my family very indignant, as they thought it eminently presumptuous, addressed to me by a man who, though a good patriot and agriculturist, knew nothing whatever about science, past or present. A good deal of political party spirit was brought into play in this instance, as is too often the case here. It is not complimentary to the state of civilisation in Italy, that in Russia and Poland, both of them very far behind her in many respects, there should exist societies for the prevention of cruelty to animals, to which all the most distinguished people have given their names.

[I rejoice to say that this stain on Italian civilisation is now wiped away. My mother just lived to hail the formation of the Società Protettrice degli Animali.—ED.

In summer we sometimes made excursions to avoid the heat of Florence. One year we went to Valombrosa and the convents of La Vernia, and Camaldoli, which are now suppressed. We travelled on mules or ponies, as the mountain paths are impracticable to carriages. I was disappointed in Valombrosa itself, but the road to it is beautiful. La Vernia is highly picturesque, there we remained two days, which I spent in drawing. The trees round the convent formed a striking contrast to the arid cliffs we had passed on the road. The monks were naturally delighted to see strangers. They belonged to the order of St. Francis, and each in his turn wandered over the country begging and living on the industry of others. We did not pay for our food and lodging, but left much more than an equivalent in the poor-box. Somerville slept in the convent, and we ladies were lodged in the so-called *Foresteria* outside; but even Somerville was not admitted into the *clausura* at Camaldoli, for the monks make a vow of perpetual silence and solitude. Each had his little separate hut and garden, and some distance above the convent, on the slopes of the Apennines, they had an establishment called the *Eremo*, for those who sought for even greater solitude. The people told us that in winter, when deep snow covers the whole place, wolves are often seen prowling about. Not far from the Eremo there is a place from whence both the Mediterranean and the Adriatic can be seen.

We occasionally went for sea-bathing to Viareggio, which is built on a flat sandy beach. The loose sand is drifted by the wind into low hillocks, and bound together by coarse grass thickly coated with silex. Among this and other plants a lovely white amaryllis, the *Pancratium Maratimum*, with a sweet and powerful perfume, springs up. We often tried to get the bulb, but it lay too deep under the sand. One evening we had gone a long way in search of these flowers, and sat down to rest, though it was beginning to be dark. We had not sat many minutes when we were surrounded by a number of what we supposed to be bats trying to get at the flowers we had gathered, but at length we discovered that they were enormous moths, which followed us home, and actually flew into the room to soar over the flowers and suck the honey with their long probosces. They were beautiful creatures with large red eyes on their wings.

* * * * *

Our life at Florence went on pretty much as usual when all at once cholera broke out of the most virulent kind. Multitudes fled from Florence; often in vain, for it prevailed all through Tuscany to a great extent. The terrified people were kneeling to the

Madonna and making processions, after which it was remarked that the number of cases was invariably increased. The Misericordia went about in their fearful costume, indefatigable in carrying the sick to the hospitals. The devotion of that society was beyond all praise; the young and the old, the artisan and the nobleman, went night and day in detachments carrying aid to the sufferers, not in Florence only, but to Fiesole and the villages round. We never were afraid, but we consulted Professor Zanetti, our medical adviser, whether we should leave the town, which we were unwilling to do, as we thought we should be far from medical assistance, and he said, "By no means; live as usual, drive out as you have always done, and make not the smallest change." We followed his advice, and drove out every afternoon till near dark, and then passed the rest of the evening with those friends who, like ourselves, had remained in town. None of us took the disease except one of our servants, who recovered from instant help being given.

The Marquis of Normanby was British minister at that time, and Lady Normanby and he were always kind and hospitable to us. At her house we became acquainted with Signora Barbieri-Nini, the celebrated opera-singer, who had retired from the stage, and lived with her husband, a Sienese gentleman, in a villa not far from Villa Normanby. She gave a musical party, to which she invited us. The music, which was entirely artistic, was excellent, the entertainment very handsome, and it was altogether very enjoyable. As we were driving home afterwards, late at night, going down the hill, our carriage ran against one of the dead carts which was carrying those who had died that day to the burying-ground at Trespiano. It was horribly ghastly—one could distinguish the forms of the limbs under the canvas thrown over the heap of dead. The burial of the poor and rich in Italy is in singular contrast; the poor are thrown into the grave without a coffin, the rich are placed in coffins, and in full dress, which, especially in the case of youth and infancy, leaves a pleasant impression. An intimate friend of ours lost an infant, and asked me to go and see it laid out. The coffin, lined with white silk, was on a table, covered with a white cloth, strewed with flowers, and with a row of wax lights on either side. The baby was clothed in a white satin frock, leaving the neck and arms bare; a rose-bud was in each hand, and a wreath of rose-buds surrounded the head, which rested on a pillow. Nothing could be prettier; it was like a sleeping angel.

* * * * *

Pio Nono had lost his popularity before he came

to visit the Grand Duke of Tuscany. The people received him respectfully, but without enthusiasm; nevertheless, Florence was illuminated in his honour. The Duomo, Campanile, and the old tower in the Piazza dei Signori were very fine, but the Lung' Arno was beautiful beyond description; the river was full, and reflected the whole with dazzling splendour.

I made the acquaintance of Signore Donati, afterwards celebrated for the discovery of one of the most brilliant comets of this century, whose course and changes I watched with the greatest interest. On one occasion I was accompanied by my valued friend Sir Henry Holland, who had come to Florence during one of his annual journeys. I had much pleasure in seeing him again.

Political parties ran very high in Florence; we sympathised with the Liberals, living on intimate terms with the chief of them. As soon as the probability of war between Piedmont and Austria became known, many young men of every rank, some even of the highest families, hastened to join as volunteers. The most sanguine long hoped that the Grand Duke might remember that he was an Italian prince rather than an Austrian archduke, and would send his troops to join the Italian cause; but his dynasty was doomed, and he blindly chose the losing side. At last the Austrians crossed the Mincio, and the war fairly broke out, France coming to the assistance of Piedmont. The enthusiasm of the Tuscans could then no longer be restrained, and on the 27th April 1859, crowds of people assembled on the Piazza dell' Indipendenza, and raised the tri-coloured flag. The government, who, the day before, had warning of what was impending, had sent sealed orders to the forts of Belvedere and del Basso, which, when opened on the eventful morning, were found to contain orders for the bombardment of the town. This the officers refused to do, after which the troops joined the popular cause. When this order became generally known, as it soon did, it proved the last blow to the dynasty, although the most eminent and respected Liberals used their best efforts during the whole of the 27th to restore harmony between the Grand Duke and the people. They advised his immediate abdication in favour of his son, the Archduke Ferdinand, the proclamation of the Constitution, and of course insisted on the immediate alliance with Piedmont as their principal condition. It was already too late! All was of no avail, and in the evening, whilst we were as usual at the Cascine, the whole Imperial family, accompanied by the Austrian minister, and escorted by several of the Corps Diplomatique, drove round the walls from Palazzo Pitti to Porta San Gallo unmolested amid a silent crowd, and crossing the frontier on the Bologna road, bade farewell for ever to Tuscany. The obnoxious ministers were also permitted to retire unnoticed to their country houses.

Thus ended this bloodless revolution; there was no disorder of any kind, which was due to the young men belonging to the principal families of Florence, such as Corsini, Incontri, Farinola, and others, using their influence with the people to calm and direct them. Indeed, so quiet was everything that my daughters walked about the streets, as did most ladies, to see what was going on; the only visible signs of the revolution throughout the whole day were bands of young men with tri-coloured flags and cockades shouting national songs at the top of their voices. As I have said already, we took our usual drive to the Cascine after dinner, and went to the theatre in the evening; the streets were perfectly quiet, and next morning the people were at work as usual. Sir James Scarlett was our minister, and had a reception the evening after these events, where we heard many predictions of evil which never were fulfilled. The least of these was the occupation of Florence by a victorious Austrian army. The Tuscan archdukes precluded all chance of a restoration by joining the Austrian army, and being present at the battle of Solferino. At Florence a provisional government was formed with Bettino Ricasoli at its head; a parliament assembled three times in the Sala dei Cinquecento, in the Palazzo Vecchio, and voted with unanimity the expulsion of the House of Lorraine, and the annexation of Tuscany to the kingdom of Italy. In the meantime the French and Italian arms were victorious in Lombardy. As, however, it is not my intention to give an historical account of the revolution of 1859, but merely to jot down such circumstances as came under my own immediate notice, I shall not enter into any particulars regarding the well-known campaign which ended in the cession of Milan and Lombardy to Italy.

We were keenly interested in the alliance between the Emperor Napoleon and the King of Italy, in hopes the Quadrilateral would be taken, and Venice added to the Italian States. We had a map of Northern Italy spread on a table, and from day to day we marked the positions of the different head-quarters with coloured-headed pins. I can hardly describe our indignation when all at once peace was signed at Villafranca, and Napoleon received Nice and Savoy in recompense for his aid, which were given up to him without regard to the will of the people. When the peace was announced in Tuscany it caused great consternation and disgust; the people were in the greatest excitement, fearing that those rulers so obnoxious to them might by this treaty be again forced upon them; and it required the firm hand of Ricasoli to calm the people, and induce the King to accept the annexation which had been voted without one dissentient voice.

Baron Ricasoli had naturally many enemies amongst the Codini, or retrograde party. Hand-grenades were thrown against the door of his house, as also at those of other ministers, but without doing harm. One evening my daughters were dressing to go to a ball that was to take place at the Palazzo delle Crocelle, close to us, in a street parallel to ours, when we were startled by a loud explosion. An attempt had been made to throw a shell into the ball-room, which had happily failed. The streets were immediately lined with soldiers, and the ball, which was given by the Ministers, as far as I recollect, took place.

When the war broke out, a large body of French troops, commanded by Prince Jerôme Napoleon, came to Florence, and were bivouacked in the Cascine. The people in the streets welcomed them as deliverers from the Austrians, whose occupation of Tuscany, when first we came to reside in Florence, was such a bitter mortification to them, and one of the causes of the unpopularity of the Grand Duke, whom they never forgave for calling in the Austrian troops after 1848. The French camp was a very pretty sight; some of the soldiers playing at games, some mending their clothes, or else cooking. They were not very particular as to what they ate, for one of my daughters saw a soldier skin a rat and put it into his soup-kettle.

We were invited by the Marchesa Lajatico, with whom we were very intimate, to go and see the entry of Victor Emmanuel into Florence from the balcony of the Casa Corsini in the Piazza del Prato, where she resides. The King was received with acclamation: never was anything like the enthusiasm. Flowers were showered down from every window, and the streets were decorated with a taste peculiar to the Italians.

[I think the following extracts from letters written by my mother during the year 1859 and the following, ever memorable in Italian history, may not be unwelcome to the reader. My mother took the keenest interest in all that occurred. Owing to the liberal opinions she had held from her youth, and to which she was ever constant, all her sympathies were with the Italian cause, and she rejoiced at every step which tended to unite all Italy in one kingdom. She lived to see this great revolution accomplished by the entry of Victor Emmanuel into Rome as King of Italy; a consummation believed by most politicians to be a wild dream of poets and hot-headed patriots, but now realised and accepted as a matter of course. My mother had always firm faith in this result, and it was with inexpressible pleasure she watched its completion. Our intimacy with the leading politicians both in Tuscany and Piedmont naturally added to our interest. Ricasoli, Menabrea, Peruzzi, Minghetti, &c., we knew intimately, as well as Camillo Cavour, the greatest statesman Italy ever produced. No one who did not witness it can imagine the grief and consternation his death occasioned, and of which my mother writes in a letter dated June 19th, 1861.

FROM MRS. SOMERVILLE TO W. GREIG, ESQ.

FLORENCE, *May 5th*, 1859.

MY DEAREST W.,

Your letter of the 28th would have made me laugh heartily were we not annoyed that you should have suffered such uneasiness on our account; the panic in England is ridiculous and most unfounded. The whole affair has been conducted with perfect unanimity and tranquillity, so that there has been no one to fight with. The Austrians are concentrated in Lombardy, and not in Tuscany, nor is there any one thing to disturb the perfect peace and quietness which prevail over the whole country; not a soul thinks of leaving Florence. You do the greatest injustice to the Tuscans. From first to last not a person has been insulted, not a cry raised against anyone; even the obnoxious ministers were allowed to go to their country houses without a word of insult, and troops were sent with the Grand Duke to escort him and his family to the frontier. Martha and Mary went all through the town the morning of the revolution, which was exactly like a common festa, and we found the tranquillity as great when we drove through the streets in the afternoon. The same quiet still prevails, the people are at their usual employments, the theatres and private receptions go on as usual, and the provisional government is excellent. Everyone knew of the revolution long before it took place and the quietness with which it was to be conducted. I am grieved at the tone of English politics, and trust, for the honour of the country and humanity, that we do not intend to make war upon France and Sardinia. It would be a disgrace and everlasting stigma to make a crusade against the oppressed, being ourselves free. The people here have behaved splendidly, and we rejoice that we have been here to witness such noble conduct. No nation ever made such progress as the Tuscans have done since the year '48. Not a word of republicanism, it has never been named. All they want is a constitutional government, and this they are quietly settling . . . . .

FROM MRS. SOMERVILLE TO W. GREIG, ESQ.

FLORENCE, *29th May*, 1859.

. . . . . . Everything is perfectly quiet here; the Tuscans are giving money liberally for carrying on the war. We have bought quantities of old linen, and your sisters and I spend the day in making lint and bandages for the wounded soldiers; great quantities have already been sent to Piedmont. Hitherto the war has been favourable to the allied army. God grant that England may not enter into the contest till the Austrians are driven out of Italy! After that point has been gained, our honour would be safe. To take part with the oppressors and maintain despotism in Italy would be infamous. Tuscany is to be occupied by a large body of troops under the command of Prince Napoleon. A great many are already encamped on the meadows at the Cascine—fine, spirited, merry young men; many of them have the Victoria medal. They are a thorough protection against any attack by the Austrians, of which, however, there is little chance, as they have enough to do in Lombardy. There is to be a great affair this morning at nine o'clock; an altar is raised in the middle of the camp, and the tricolour (Italian) flag is to be blessed amidst salvoes of cannon. Your friend, Bettino Ricasoli, is thought by far the most able and statesmanlike person in Tuscany; he is highly respected. Martha and I dined with Mr. Scarlett, and met . . . who said if the Grand Duke had not been the most foolish and obstinately weak man in the world, he might still have been on the throne of Tuscany; but that now he has made that impossible by going to Vienna and allowing his two sons to enter the Austrian army. . . . . We have had a visit from Dr. Falkner, his two nieces and brother. They had been spending the winter in Sicily, where he discovered rude implements formed by man mixed with the bones of prehistoric animals in a cave, so hermetically shut up that not a doubt is left of a race of men having lived at a period far anterior to that assigned as the origin of mankind. Similar discoveries have recently been made elsewhere. Dr. Falkner had travelled much in the Himalayas, and lived two years on the great plain of Tibet; the account he gave me of it was most interesting. His brother had spent fifteen years in Australia, so the conversation delighted me; I learnt so much that was new. I am glad to hear that the Queen has been so kind to my friend Faraday; it seems she has given him an apartment at Hampton Court nicely fitted up. She went to see it herself, and having consulted scientific men as to the instruments that were necessary for his pursuits, she had a laboratory fitted up with them, and made him a present of the whole. That is doing things handsomely, and no once since Newton has deserved it so much.

FROM MRS. SOMERVILLE TO W. GREIG, ESQ.

FLORENCE, *5th June*, 1859.

. . . . . . All is perfectly quiet; a large body of French troops are now in Tuscany, and many more are expected probably to make a diversion on this side of the Austrian army through Modena; but nothing is known; the most profound secrecy is maintained as to all military movements. Success has hitherto attended the allied army, and the greatest bravery has been shown. The enthusiasm among the men engaged is excessive, the King of Sardinia himself the bravest of the brave, but exposes himself so much that the people are making petitions to him to be more careful. The Zouaves called out in the midst of the battle, "Le roi est un Zouave!" Prince Napoleon keeps very quiet, and avoids shewing himself as much as possible. The French troops are very fine indeed—young, gay, extremely civil and well bred. The secrecy is quite curious; even the colonels of the regiments do not know where they may be snet till the order comes: so all is conjecture. . . . . . . The young King of Naples seems to follow the footsteps of his father; I hope in God that we may not protect and defend him. How anxious we are to know what the House of Commons will do! Let us hope they will take the liberal side; but the conservative party seems to be increasing.

FROM MRS. SOMERVILLE TO W. GREIG, ESQ.

FLORENCE, *22nd August*, 1859.

. . . . . . Public affairs go on admirably. A few weeks ago the elections took place of the members of the Tuscan parliament with a calm and tranquillity of which you have no idea. Every proprietor who pays 15 pauls of taxes (75 pence) has a vote. There are 180 members, consisting of the most ancient nobility, the richest proprietors, the most distinguished physicians and lawyers, and the most respectable merchants. They hold their meetings in the magnificent hall of the Palazzo Vecchio—the Sala Dei Cinquecento. The first two or three days were employed in choosing a president &c., &c.; then a day was named to determine the fate of the house of Lorraine. I could not go, but Martha went with a Tuscan friend. There was no speaking; the vote was by ballot, and each member separately went up to a table before the president, and silently put his ball into a large vase. Two members poured the balls into a tray, and on examination, said, "No division is necessary; they are *all* black,"—which was followed by long and loud cheering. They have been equally unanimous in the Legations in Parma and Modena; and the wish of the people is to form one kingdom of these four states under an Italian prince, excluding all Austrians for ever. The union is perfect, and the determination quiet but deep and unalterable. If the Archduke is forced upon them, it must be by armed force, which the French emperor will not likely permit, after the Archduke was fool enough to fight against him at Solferino. All the four states have unanimously voted union with Piedmont; but they do not expect it to be granted. The destinies of Europe are now dependent on the two emperors. . . . .

FROM MRS. SOMERVILLE TO W. GREIG, ESQ.

FLORENCE, 23*rd April*, 1860.

You would have had this letter sooner, my dearest Woronzow, if I had not been prevented from writing to you yesterday evening. . . . . . The weather has been atrocious; deluges of rain night and day, and so cold that I have been obliged to lay in a second supply of wood. The only good day, and the only one I have been out, was that on which the king arrived. It fortunately was fine, and the sight was magnificent; quite worthy of so great an historical event. No carriages were allowed after the guns fired announcing that the king had left Leghorn; so we should have been ill off, had it not been for the kindness of our friend the Marchesa Lajatico, who invited us to her balcony, which is now very large, as they have built an addition to their house for the eldest son and his pretty wife. We were there some hours before the king arrived; but as all the Florentine society was there, and many of our friends from Turin and Genoa, we found it very agreeable. The house is in the Prato, very near the gate the king was to enter. On each side of it stages were raised like steps in an amphitheatre, which were densely crowded, every window decorated with gaily-coloured hangings and the Italian flag; the streets were lined with "guardie civiche," and bands of music played from time to time. The people shouted "Evviva!" every time a gun was fired. In the midst of this joy, there appeared what resembled a funeral procession—about a hundred emigrants following the Venetian, Roman, and Neapolitan colours, all hung with black crape; they were warmly applauded, and many people shed tears. They went to the railway station just without the gate to meet the King, and when they hailed him as "*Rè d'Italia!*" he was much affected. At last he appeared riding a fine English horse, Prince Carignan on one hand and Baron Ricasoli on his left, followed by a numerous "*troupe dorée*" of generals and of his suite in gay uniforms and well mounted. The King rides well; so the effect was extremely brilliant. Then followed several carriages; in the first were Count Cavour, Buoncompagni, and the Marchese Bartolommei. You cannot form the slightest idea of the excitement; it was a burst of enthusiasm, and the reception of Cavour was as warm. We threw a perfect shower of flowers over him, which the Marchesa had provided for the occasion; and her youngest son Cino, a nice lad, went himself to present his bouquet to the King, who seemed quite pleased with the boy. I felt so much for Madame de Lajatico herself. . . . . . . I said to her how kind I thought it in her to open her house; she burst into tears, and said, though she was in deep affliction, she could not be so selfish as not offer her friends the best position in Florence for seeing what to many of them was the most important event in their lives, as it was to her even in her grief. The true Italian taste appeared to perfection in every street through which the procession passed to the Duomo, and thence to the Palazzo Pitti. Those who saw it declare nothing could surpass the splendour of the cathedral when illuminated; but that we could not see, nor did we see the procession again; it was impossible to penetrate the crowd. They say there are 40,000 strangers in Florence. . . . . . . I was much too tired to go out again to see the illuminations and the fireworks on the Ponte Carraja; your sisters saw it all, so I leave them to tell you all about it. The King and Prince are terribly early; they and Ricasoli are on horseback by *five* in the morniug; the King dines at twelve, and never touches food afterwards, though he has a dinner party of 60 or 80 every day at six. . . . . . . Now, my dearest Woronzow, I must end, for I do not wish to miss another post. I am really wonderfully well for my age.

Your devoted mother,

MARY SOMERVILLE.

FROM MRS. SOMERVILLE TO W. GREIG, ESQ.

FLORENCE, 19*th June*, 1861.

. . . . . Italy has been thrown into the deepest affliction by the death of Cavour. In my long life I never knew any event whatever which caused so universal and deep sorrow. There is not a village or town throughout the whole peninsula which has not had a funeral service, and the very poorest people, who had hardly clothes on their backs, had black crape tied round their arm or neck.

It was a state of consternation, and no wonder! Every one felt that the greatest and best man of this century has been taken away before he had completely emancipated his country. All the progress is due to him, and to him alone; the revolution has called forth men of much talent, yet the whole are immeasurably his inferior in every respect—even your friend, Ricasoli, who is most able, and the best successor that can be found, is, compared with Cavour, as Tuscany to Europe. Happily the sad loss did not occur sooner. Now things are so far advanced that they cannot go back, and I trust that Ricasoli, who is not wanting in firmness and moral courage, will complete what has been so happily begun. I am sorry to say he is not in very good health, but I trust he will not fall into the hands of the physician who attended Cavour, and who mistook his disease, reduced him by loss of blood, and then finding out his real illness, tried to strengthen him when too late. There was a most excellent article in the "Times" on the two statesmen.

[My mother's recollections continue thus:—

One night the moon shone so bright that we sent the carriage away, and walked home from a reception at the Marchesa Ginori's. In crossing the Piazza San Marco, an acquaintance, who accompanied us, took us to the Maglio, which is close by, to hear an echo. I like an echo; yet there is something so unearthly in the aërial voice, that it never fails to raise a superstitious chill in me, such as I have felt more than once as I read "Ossian" while travelling among our Highland hills in my early youth. In one of the grand passes of the Oberland, when we were in Switzerland, we were enveloped in a mist, through which peaks were dimly seen. We stopped to hear an echo; the response came clear and distinct from a great distance, and I felt as if the Spirit of the Mountain had spoken. The impression depends on accessory circumstances; for the roar of a railway train passing over a viaduct has no such effect.

* * * * *

I lost my husband in Florence on the 26th June, 1860. . . . . From the preceding narrative may be seen the sympathy, affection, and confidence, which always existed between us. . . . .

[After what has already been said of the happiness my mother enjoyed during the long years of their married life, it may be imagined what grief was her's at my father's death after only three days' illness. My mother's dear friend and correspondent, Miss F. P. Cobbe, wrote to her as follows on this occasion:—

"I have just learned from a letter from Captain Fairfax to my brother the great affliction which has befallen you. I cannot express to you how it has grieved me to think that such a sorrow should have fallen on you, and that the dear, kind old man, whose welcome so often touched and gratified me, should have passed away so soon after

I had seen you both, as I often thought, the most beautiful instance of united old age. His love and pride in you, breaking out as it did at every instant when you happened to be absent, gives me the measure of what his loss must be to your warm heart."

The following letter from my mother, dated April, 1861, addressed to her sister-in-law, was written after reading my grandfather's "Life and Times," the publication of which my father did not live to see.

FROM MRS. SOMERVILLE TO MRS. ELLIOT, OF ROSEBANK, ROXBURGHSHIRE.

FLORENCE, 28*th April*, 1861.

MY DEAR JANET,—

I received the precious volume* you have so kindly sent to me some days ago, but I have delayed thanking you for it till now because we all wished to read it first. We are highly pleased, and have been deeply interested in it. The whole tone of the book is characteristic of your dear father; the benevolence, warm-heartedness, and Christian charity which appeared in the whole course of his life and ministry. That which has struck us all most forcibly is the liberality of his sentiments, both religious and political, at a time when narrow views and bigotry made it even dangerous to avow them, and it required no small courage to do so. He was far in advance of the age in which he lived;

* The Rev. P. Somerville's "Life and Times."

his political opinions are those of the present day, his religious opinions still before it. There are many parts of the book which will please the general reader from the graphic description of the manners and customs of the time, as well as the narrative of his intercourse with many of the eminent men of his day. Your most dear father's affectionate remembrance of me touches me deeply. I have but one regret, dear Jenny, and that is that our dear William did not live to see the accomplishment of what was his dying wish; but God's will be done. . . . . We are all much as usual: I am wonderfully well, and able to write, which I do for a time every day. I do not think I feel any difference in capacity, but I become soon tired, and then I read the newspapers, some amusing book, or work. . . . Everything is flourishing in Italy, and the people happy and contented, except those who were employed and dependent on the former sovereigns, but they are few in comparison; and now there is a fine army of 200,000 men to defend the country, even if Austria should make an attack, but that is not likely at present. Rome is still the difficulty, but the Pope must and soon will lose his temporal power, for the people are determined it shall be so. . . .

I am, dear sister,

Most affectionately yours,

MARY SOMERVILLE.

TO MRS. ELLIOT, of Rosebank, Roxburghshire.

CHAPTER XVII.

SPEZIA—GENOA—BEGINS MOLECULAR AND MICROSCOPIC SCIENCE—TURIN—SPEZIA—BRITISH FLEET—LETTERS FROM MRS. SOMERVILLE—GARIBALDI—SEVERE ILLNESS—FLORENCE—MY BROTHER'S DEATH—NAPLES—ERUPTION OF VESUVIUS—J. S. MILL—CHANGE IN PUBLIC OPINION ON WOMEN'S EDUCATION—EIGHTY-NINTH YEAR—DESCRIBES HER OWN CHARACTER—THOUGHTS ON A FUTURE LIFE—PROGRESS IN KNOWLEDGE OF GEOGRAPHY—VICTORIA MEDAL—MEDAL FROM ROYAL ITALIAN GEOGRAPHICAL SOCIETY—LETTER FROM MENABREA—ROME, CAPITAL OF ITALY—AURORA BOREALIS.

SOON after my dear husband's death, we went to Spezia, as my health required change, and for some time we made it our headquarters, spending one winter at Florence, another at Genoa, where my son and his wife came to meet us, and where I had very great delight in the beautiful singing of our old friend Clara Novello, now Countess Gigliucci, who used to come to my house, and sing Handel to me. It was a real pleasure, and her voice was as pure and silvery as when I first heard her, years before. Another winter we spent at Turin. On returning to Spezia in the summer of 1861, the beautiful comet visible that year appeared for the first time the very evening we arrived. On the following, and during many evenings while it was visible, we used to row in a small boat a little way from shore, in order to see it to greater advantage. Nothing could be more poetical than the clear starlit heavens with this beautiful comet reflected, nay, almost repeated, in the calm glassy water of the gulf. The perfect silence and stillness of the scene was very impressive.

I was now unoccupied, and felt the necessity of having something to do, desultory reading being insufficient to interest me; and as I had always considered the section on chemistry the weakest part of the connection of the "Physical Sciences," I resolved to write it anew. My daughters strongly opposed this, saying, "Why not write a new book?" They were right; it would have been lost time: so I followed their advice, though it was a formidable undertaking at my age, considering that the general character of science had greatly changed. By the improved state of the microscope, an invisible creation in the air, the earth, and the water, had been brought within the limits of human vision; the microscopic structure of plants and animals had been minutely studied, and by synthesis many substances had been formed of the elementary atoms similar to those produced by nature. Dr. Tyndall's experiments had proved the inconceivable minuteness of the atoms of matter; M. Gassiot and Professor Plücher had published their experiments on the stratification of the electric light; and that series of discoveries by scientific men abroad, but chiefly by our own philosophers at home, which had been in progress for a course of years, prepared the way for Bunsen and Kirchhof's marvellous consummation.

Such was the field opened to me; but instead of being discouraged by its magnitude, I seemed to have resumed the perseverance and energy of my youth, and began to write with courage, though I did not think I should live to finish even the sketch I had made, and which I intended to publish under the name of "Molecular and Microscopic Science," and assumed as my motto, "Deus magnus in magnis, maximus in minimis," from Saint Augustin.

My manuscript notes on Science were now of the greatest use; and we went for the winter to Turin (1861—1862), where I could get books from the public libraries, and much information on subjects of natural history from Professor De Filippi, who has recently died, much regretted, while on a scientific mission to Japan and China, as well as from other sources. I subscribed to various periodicals on chemical and other branches of science; the transactions of several of our societies were sent to me, and I began to write. I was now an old woman, very deaf and with shaking hands; but I could still see to thread the finest needle, and read the finest print, but I got sooner tired when writing than I used to do. I wrote regularly every morning from eight till twelve or one o'clock before rising. I was not alone, for I had a mountain sparrow, a great pet, which sat, and indeed is sitting on my arm as I write these lines.

The Marchese Doria has a large property at Spezia, and my dear friend Teresa Doria generally spent the evening with us, when she and I chatted and played Bézique together. Her sons also came frequently, and some of the officers of the Italian navy. One who became our very good friend is Captain William Acton, now Admiral, and for two years Minister of Marine; he is very handsome, and, what is better, a most agreeable, accomplished gentleman, who has interested himself in many branches of natural history, besides being a good linguist. In summer the British squadron, commanded by Admiral Smart, came for five weeks to Spezia. My nephew, Henry Fairfax, was commander on board the ironclad "Resistance." Notwithstanding my age, I was so curious to see an ironclad that I went all over the "Resistance," even to the engine-room and screw-alley. I also went to luncheon on board the flag-ship "Victoria," a three-decker, which put me in mind of olden times.

[The following extracts are from letters of my mother's, written in 1863 and 1865:—

FROM MRS. SOMERVILLE TO W. GREIG, ESQ.

SPEZIA, *12th May*, 1863.

How happy your last letter has made me, my dearest Woronzow, to hear that you are making real progress, and that you begin to feel better from the Bath waters . . . . . Of your general health I had the very best account this morning from your friend Colonel Gordon. I was most agreeably surprised and gratified by a very kind and interesting letter from him, enclosing his photograph, and giving me an account of his great works at Portsmouth with reference to the defence by iron as well as stone . . . . . .

I wish I could show you the baskets full of flowers which Martha and Mary bring to me from the mountains. They are wonderfully beautiful; it is one of my greatest amusements putting them in water. I quite regret when they cannot go for them. The orchises and the gladioles are the chief flowers now, but such a variety and such colours! You see we have our quiet pleasures. I often think of more than "60 years ago," when I used to scramble over the Bin at Burntisland after our tods-tails and leddies-fingers, but I fear there is hardly a wild spot existing now in the lowlands of Scotland . . . . . .

God bless you, my dearest Woronzow.

FROM MRS. SOMERVILLE TO W. GREIG, ESQ.

SPEZIA, 27*th Sept.*, 1865.

MY DEAREST WORONZOW,

I fear Agnes and you must have thought your old mother had gone mad when you read M.'s letter. In my sober senses, however, though sufficiently excited to give me strength for the time, I went over every part of the *Resistance*,* and examined everything in detail except the *stokehole!* I was not even hoisted on board, but mounted the companion-ladder bravely. It was a glorious sight, the perfection of structure in every part astonished me. A ship like that is the triumph of human talent and of British talent, for all confess our superiority in this respect to every other nation, and I am happy to see that no jealousy has arisen from the meeting of the French and English fleets. I was proud that our "young admiral" † had the command of so fine a vessel . . . . I also spent a most agreeable day on board the *Victoria*, three-decker, and saw every part of the three decks, which are very different from what they were in my father's time; everything on a much larger scale, more elegant and convenient. But the greatest change is in the men; I never saw a finer set, so gentlemanly-looking and well-behaved; almost all can read and write, and they have an excellent library and reading-room in all the ships. No sooner was the fleet gone than the Italian Society of Natural History held their annual

* The *Resistance*, ironclad, commanded by Captain Chamberlayne, then absent on sick leave.

† Captain Henry Fairfax, my mother's nephew, then Commander on board the *Resistance*, senior officer in the absence of the captain.

meeting here, Capellini* being president in the absence (in Borneo) of Giacomo Doria. There were altogether seventy members, Italian, French, and German. I was chosen an Associate by acclamation, and had to write a few lines of thanks. The weather was beautiful and the whole party dined every day on the terrace below our windows, which was very amusing to Miss Campbell and your sisters, who distinctly heard the speeches. I was invited to dinner and the wife of the celebrated Professor Vogt was asked to meet me; I declined dining, as it lasted so long that I should have been too tired, but I went down to the dessert. Capellini came for me, and all rose as I came in, and every attention was shown me, my health was drank, &c. &c. It lasted four days, and we had many evening visits, and I received a quantity of papers on all subjects. I am working very hard (for me at least), but I cannot hurry, nor do I see the need for it. I write so slowly on account of the shaking of my hand that although my head is clear I make little but steady progress . . . . . .

Your affectionate mother,

MARY SOMERVILLE.

After the battle of Aspromonte, Garibaldi arrived a prisoner on board a man-of-war, and was placed at Varignano under surveillance. His wound had not been properly dressed, and he was in a state of great suffering. Many surgeons came from all parts of Italy, and one even from England, to attend him, but the

* Professor of Geology at Bologna.

eminent Professor Nélaton saved him from amputation, with which he was threatened, by extracting the bullet from his ankle. I never saw Garibaldi during his three months' residence at Varignano and Spezia; I had no previous acquaintance with him; consequently, as I could be of no use to him, I did not consider myself entitled to intrude upon him merely to gratify my own curiosity, although no one admired his noble and disinterested character more than I did. Not so, many of my countrymen, and countrywomen too, as well as ladies of other nations, who worried the poor man out of his life, and made themselves eminently ridiculous. One lady went so far as to collect the hairs from his comb,—others showered tracts upon him.

I had hitherto been very healthy; but in the beginning of winter I was seized with a severe illness which, though not immediately dangerous, lasted so long, that it was doubtful whether I should have stamina to recover. It was a painful and fatiguing time to my daughters. They were quite worn out with nursing me; our maid was ill, and our man-servant, Luigi Lucchesi, watched me with such devotion that he sat up twenty-four nights with me. He has been with us eighteen years, and now that I am old and feeble, he attends me with unceasing kindness. It is but justice to say that we never were so faithfully or well served as by Italians; and none are more ingenious in turning their hands to anything, and in never objecting to do this or that, as not what they were hired for,—a great quality for people who, like ourselves, keep few servants. After a time they identify themselves with the family they serve, as my faithful Luigi has done with all his heart. I am sincerely attached to him.

* * * * *

In the spring, when I had recovered, my son and his wife came to Spezia, and we all went to Florence, where we had the pleasure of seeing many old friends. We returned to Spezia, and my son and his wife left us to go back to England, intending to meet us again somewhere the following spring. I little thought we never should meet again. . . . . . My son sent his sisters a beautiful little cutter, built by Mr. Forrest in London, which has been a great resource to them. I always insist on their taking a good sailor with them, although I am not in the least nervous for their safety. Indeed, small as the "Frolic" is—and she is only about twenty-eight feet from stem to stern—she has weathered some stiff gales gallantly, as, for instance, when our friend, Mr. Montague Brown, British consul at Genoa, sailed her from Genoa to Spezia in very bad weather; and in a very dangerous squall my daughters were caught in, coming from Amalfi to Sorrento. The "Frolic" had only just arrived at Spezia, when we heard of the sudden death of my dear son, Oct., 1865.

[This event, which took from my mother's last years one of her chief delights, she bore with her usual calm courage, looking forward confidently to a reunion at no distant date with one who had been the most dutiful of sons and beloved of friends. She never permitted herself, in writing her Recollections, to refer to her feelings under these great sorrows.

Some time after this, my widowed daughter-in-law spent a few months with us. On her return to London, I sent the manuscript of the "Molecular and Microscopic Science" with her for publication. In writing this book I made a great mistake, and repent it. Mathematics are the natural bent of my mind. If I had devoted myself exclusively to that study, I might probably have written something useful, as a new era had begun in that science. Although I got "Chales on the Higher Geometry," it could be but a secondary object while I was engaged in writing a popular book. Subsequently, it became a source of deep interest and occupation to me.

Spezia is very much spoilt by the works in progress for the arsenal, though nothing can change the beauty of the gulf as seen from our windows, especially the group of the Carrara mountains, with fine peaks and ranges of hills, becoming more and more verdant down to the water's edge. The effect of the setting-sun on this group is varied and brilliant beyond belief. Even I, in spite of my shaking hand, resumed the brush, and painted a view of the ruined Castle of Ostia, at the mouth of the Tiber, from a sketch of my own, for my dear friend Teresa Doria.

We now came to live at Naples; and on leaving Spezia, I spent a fortnight with Count and Countess Usedom at the Villa Capponi, near Florence, where, though unable to visit, I had the pleasure of seeing my Florentine friends again.

We spent two days in Rome, and dined with our friends the Duca and Duchesa di Sermoneta. We were grieved at his blindness, but found him as agreeable as ever.

Through our friend, Admiral Acton, I became acquainted with Professor Panceri, Professor of Comparative Anatomy; Signore de Gasparis, who has discovered nine of the minor planets, and is an excellent mathematician, and some others. To these gentlemen I am indebted for being elected an honorary member of the Accademia Pontoniana.

We were much interested in Vesuvius, which, for several months, was in a state of great activity. At first, there were only volumes of smoke and some small streams of lava, but these were followed by the most magnificent projections of red hot stones and rocks rising 2,000 feet above the top of the mountain. Many fell back again into the crater, but a large portion were thrown in fiery showers down the sides of the cone. At length, these beautiful eruptions of *lapilli* ceased, and the lava flowed more abundantly, though, being intermittent and always issuing from the summit, it was quite harmless; volumes of smoke and vapour rose from the crater, and were carried by the wind to a great distance. In sunshine the contrast was beautiful, between the jet-black smoke and the silvery-white clouds of vapour. At length, the mountain returned to apparent tranquillity, though the violent detonations occasionally heard gave warning that the calm might not last long. At last, one evening, in November, 1868, when one of my daughters and I were observing the mountain through a very good telescope, lent us by a friend, we distinctly saw a new crater burst out at the foot of the cone in the Atrio del Cavallo, and bursts of red-hot lapilli and red smoke pouring forth in volumes. Early next morning we saw a great stream of lava pouring down to the north of the Observatory, and a column of black smoke issuing from the new craters, because there were two, and assuming the well-known appearance of a pine-tree. The trees on the northern edge of the lava were already on fire. The stream of lava very soon reached the plain, where it overwhelmed fields, vineyards, and houses. It was more than a mile in width and thirty feet deep. My daughters went up the mountain the evening after the new craters were formed; as for me, I could not risk the fatigue of such an excursion, but I saw it admirably from our own windows. During this year the volcanic forces in the interior of the earth were in unusual activity, for a series of earthquakes shook the west coast of South America for more than 2,500 miles, by which many thousands of the inhabitants perished, and many more were rendered homeless. Slight shocks were felt in many parts of Europe, and even in England. Vesuvius was our safety-valve. The pressure must have been very great which opened two new craters in the Atrio del Cavallo and forced out such a mass of matter. There is no evidence that water had been concerned in the late eruption of Vesuvius; but during the whole of the preceding autumn, the fall of rain had been unusually great and continuous. There were frequent thunder-storms ; and, on one occasion, the quantity of rain that fell was so great, as to cause a land-slip in Pizzifalcone, by which several houses were overwhelmed ; and, on another occasion, the torrent of rain was so violent, that the Riviera di Chiaja was covered, to the depth of half a metre, with mud, and stones brought down by the water from the heights above. This enormous quantity of water pouring on the slopes of Vesuvius, and percolating through the crust of the earth into the fiery caverns, where volcanic forces are generated, being resolved into steam, and possibly aided by the expansion of volcanic gases, may have been a partial agent in propelling the formidable stream of lava which has caused such destruction. We observed, that when lava abounded, the projection of rocks and lapilli either ceased altogether, or became of small amount. The whole eruption ended in a shower of impalpable ashes, which hid the mountain for many days, and which were carried to a great distance by the wind. Sometimes the ashes were pure white, giving the mountain the appearance of being covered with snow. Vapour continued to rise from Vesuvius in beautiful silvery clouds, which ceased and left the edge of the crater white with sublimations. I owe to Vesuvius the great pleasure of making the acquaintance of Mr. Phillips, Professor of Geology in the University of Oxford; and, afterwards, that of Sir John Lubbock, and Professor Tyndall, who had come to Naples on purpose to see the eruption. Unfortunately, Sir John Lubbock and Professor Tyndall were so limited for time, that they could only spend one evening with us; but I enjoyed a delightful evening, and had much scientific conversation.

Notwithstanding the progress meteorology has made since it became a subject of exact observation, yet no explanation has been given of the almost unprecedented high summer temperature of 1868 in Great Britain, and even in the Arctic regions. In England, the grass and heather were dried up, and extensive areas were set on fire by sparks from railway locomotives, the conflagrations spreading so rapidly, that they could only be arrested by cutting trenches to intercept their course. The whalers found open water to a higher latitude than usual; but, although the British Government did not avail themselves of this opportunity for further Arctic discovery, Sweden, Germany, France, and especially the United States, have taken up the subject with great energy. Eight expeditions sailed for the North Polar region between the years 1868 and 1870; several for the express purpose of reaching the

Polar Sea, which, I have no doubt, will be attained, now that steam has given such power to penetrate the fields of floating ice. It would be more than a dashing exploit to make a cruise on that unknown sea; it would be a discovery of vast scientific importance with regard to geography, magnetism, temperature, the general circulation of the atmosphere and oceans, as well as to natural history. I cannot but regret that I shall not live to hear the result of these voyages.

* * * * *

The British laws are adverse to women; and we are deeply indebted to Mr. Stuart Mill for daring to show their iniquity and injustice. The law in the United States is in some respects even worse, insulting the sex, by granting suffrage to the newly-emancipated slaves, and refusing it to the most highly-educated women of the Republic.

[For the noble character and transcendent intellect of Mr. J. S. Mill my mother had the greatest admiration. She had some correspondence with him on the subject of the petition to Parliament for the extension of the suffrage to women, which she signed; and she also wrote to thank him warmly for his book on the "Subjection of Women." In Mr. Mill's reply to the latter he says:—

FROM JOHN STUART MILL, ESQ., TO MRS. SOMERVILLE.

BLACKHEATH PARK, *July 12th*, 1869.

DEAR MADAM,

Such a letter as yours is a sufficient reward for the trouble of writing the little book. I could have desired no better proof that it was adapted to its purpose than such an encouraging opinion from you. I thank you heartily for taking the trouble to express, in such kind terms, your approbation of the book,—the approbation of one who has rendered such inestimable service to the cause of women by affording in her own person so high an example of their intellectual capabilities, and, finally, by giving to the protest in the great Petition of last year the weight and importance derived from the signature which headed it.

I am,

Dear Madam,

Most sincerely and respectfully yours,

J. S. MILL.

Age has not abated my zeal for the emancipation of my sex from the unreasonable prejudice too prevalent in Great Britain against a literary and scientific education for women. The French are more civilized in this respect, for they have taken the lead, and have given the first example in modern times of encouragement to the high intellectual culture of the sex. Madame Emma Chenu, who had received the degree of Master of Arts from the Academy of Sciences in Paris, has more recently received the diploma of Licentiate in Mathematical Sciences from the same illustrious Society, after a successful examination in algebra, trigonometry, analytical geometry, the differential and integral calculi, and astronomy. A Russian lady has also taken a degree; and a lady of my acquaintance has received a gold medal from the same Institution.

I joined in a petition to the Senate of London University, praying that degrees might be granted to women; but it was rejected. I have also frequently signed petitions to Parliament for the Female Suffrage, and have the honour now to be a member of the General Committee for Woman Suffrage in London.

* * * * *

[My mother, in alluding to the great changes in public opinion which she had lived to see, used to remark that a commonly well-informed woman of the present day would have been looked upon as a prodigy of learning in her youth, and that even till quite lately many considered that if women were to receive the solid education men enjoy, they would forfeit much of their feminine grace and become unfit to perform their domestic duties. My mother herself was one of the brightest examples of the fallacy of this old-world theory, for no one was more thoroughly and gracefully feminine than she was, both in manner and appearance; and, as I have already mentioned, no amount of scientific labour ever induced her to neglect her home duties. She took the liveliest interest in all that has been done of late years to extend high class education to women, both classical and scientific, and hailed the establishment of the Ladies' College at Girton as a great step in the true direction, and one which could not fail to obtain most important results. Her scientific library, as already stated, has been presented to this College as the best fulfilment of her wishes.

* * * * *

I have lately entered my 89th year, grateful to God for the innumerable blessings He has bestowed on me and my children; at peace with all on earth, and I trust that I may be at peace with my Maker when my last hour comes, which cannot now be far distant.

Although I have been tried by many severe afflictions, my life upon the whole has been happy. In my youth I had to contend with prejudice and illiberality; yet I was of a quiet temper, and easy to live with, and I never interfered with or pryed into other people's affairs. However, if irritated by what I considered unjust criticism or interference with myself, or any one I loved, I could resent it fiercely. I was not good at argument; I was apt to lose my temper; but I never bore ill will to any one, or forgot the manners of a gentlewoman, however angry I may have been at the time. But I must say that no one ever met with such kindness as I have done. I never had an enemy. I have never been of a melancholy disposition; though depressed sometimes by circumstances, I always rallied again; and although I seldom laugh, I can laugh heartily at wit or on fit occasion. The short time I have to live naturally occupies my thoughts. In the blessed hope of meeting again with my beloved children, and those who were and are dear to me on earth, I think of death with composure and perfect confidence in the mercy of God. Yet to me, who am afraid to sleep alone on a stormy night, or even to sleep comfortably any night unless some one is near, it is a fearful thought, that my spirit must enter that new state of existence quite alone. We are told of the infinite glories of that state, and I believe in them, though it is incomprehensible to us; but as I do comprehend, in some degree at least, the exquisite loveliness of the visible world, I confess I shall be sorry to leave it. I shall regret the sky, the sea, with all the changes of their beautiful colouring; the earth, with its verdure and flowers: but far more shall I grieve to leave animals who have followed our steps affectionately for years, without knowing for certainty their ultimate fate, though I firmly believe that the living principle is never extinguished. Since the atoms of matter are indestructible, as far as we know, it is difficult to believe that the spark which gives to their union life, memory, affection, intelligence, and fidelity, is evanescent. Every atom in the human frame, as well as in that of animals, undergoes a periodical change by continual waste and renovation; the abode is changed, not its inhabitant. If animals have no future, the existence of many is most wretched; multitudes are starved, cruelly beaten, and loaded during life; many die under a barbarous vivisection. I cannot believe that any creature was created for uncompensated misery; it would be contrary to the attributes of God's mercy and justice. I am sincerely happy to find that I am not the only believer in the immortality of the lower animals.

* * * * *

When I was taught geography by the village schoolmaster at Burntisland, it seemed to me that half the world was *terra incognita*, and now that a new edition of my "Physical Geography" is required, it will be a work of great labour to bring it up to the present time. The discoveries in South Africa alone would fill a volume. Japan and China have been opened to Europeans since my last edition. The great continent of Australia was an entirely unknown country, except part of the coast. Now telegrams have been sent and answers received in the course of a few hours, from our countrymen throughout that mighty empire, and even from New Zealand, round half the globe. The inhabitants of the United States are our offspring ; so whatever may happen to Great Britain in the course of events, it still will have the honour of colonizing, and consequently civilizing, half the world.

In all recent geographical discoveries, our Royal Geographical Society has borne the most important part, and none of its members have done more than my highly-gifted friend the President, Sir Roderick Murchison, geologist of Russia, and founder and author of the colossal "Silurian System." To the affection of this friend, sanctioned by the unanimous approval of the council of that illustrious Society, I owe the honour of being awarded the Victoria Medal for my "Physical Geography." An honour so unexpected, and so far beyond my merit, surprised and affected me more deeply than I can find words to express.

In the events of my life it may be seen how much I have been honoured by the scientific societies and universities of Italy, many of whom have elected me an honorary member or associate ; but the greatest honour I have received in Italy has been the gift of the first gold medal hitherto awarded by the Geographical Society at Florence, and which was coined on purpose, with my name on the reverse. I received it the other day, accompanied by the following letter from General Menabrea, President of the Council, himself a distinguished mathematician and philosopher:—

FROM GENERAL MENABREA TO MRS. SOMERVILLE.

FLORENCE, 30 *Juin*, 1869.

MADAME,

J'ai pris connaissance avec le plus grand intérêt de la belle édition de votre dernier ouvrage sur la Géographie Physique, et je désire vous donner un témoignage d'haute estime pour vos travaux. Je vous prie donc, Madame, d'accepter une médaille d'or à l'effigie du Roi Victor Emmanuel, mon auguste souverain. C'est un souvenir de mon pays dans lequel vous comptez, comme chez toutes les nations où la science est honoré, de nombreux amis et admirateurs. Veuillez croire, Madame, que je ne cesserai d'être l'un et l'autre en même temps que je suis,

Votre très dévoué Serviteur,

MENABREA.

At a general assembly of the Italian Geographical Society, at Florence, on the 14th March, 1870, I was elected by acclamation an Honorary Associate of that distinguished society. I am indebted to the President, the Commendatore Negri, for having proposed my name, and for a very kind letter, informing me of the honour conferred upon me.

* * * * *

I have still (in 1869) the habit of studying in bed from eight in the morning till twelve or one o'clock; but, I am left solitary; for I have lost my little bird who was my constant companion for eight years. It had both memory and intelligence, and such confidence in me as to sleep upon my arm while I was writing. My daughter, to whom it was much attached, coming into my room early, was alarmed at its not flying to meet her, as it generally did, and at last, after a long search, the poor little creature was found drowned in the jug.

On the 4th October, while at dinner, we had a shock of earthquake. The vibrations were nearly north and south; it lasted but a few seconds, and was very slight; but in Calabria, &c., many villages and towns were overthrown, and very many people perished. The shocks were repeated again and again; only one was felt at Naples; but as it occurred in the night, we were unconscious of it. At Naples, it was believed there would be an eruption of Vesuvius; for the smoke was particularly dense and black, and some of the wells were dried up.

* * * * *

I can scarcely believe that Rome, where I have spent so many happy years, is now the capital of united Italy. I heartily rejoice in that glorious termination to the vicissitudes the country has undergone, and only regret that age and infirmity prevent me from going to see Victor Emmanuel triumphantly enter the capital of his kingdom. The Pope's reliance on foreign troops for his safety was an unpardonable insult to his countrymen.

* * * * *

The month of October this year (1870), seems to have been remarkable for displays of the Aurora Borealis. It seriously interfered with the working of the telegraphs, particularly in the north of England and Ireland. On the night of the 24th October, it was seen over the greater part of Europe. At Florence, the common people were greatly alarmed, and at Naples, the peasantry were on their knees to the Madonna to avert the evil. Unfortunately, neither I nor any of my family saw the Aurora; for most of our windows have a southern aspect. The frequent occurrence of the Aurora in 1870 confirms the already known period of maximum intensity and frequency, every ten or twelve years, since the last maximum occurred in 1859.

CHAPTER XVIII.

ECLIPSE—VISITS OF SCIENTIFIC MEN—LIFE AT NAPLES—DARWIN'S BOOKS —REMARKS ON CIVILIZATION—FINE AURORA BOREALIS—DEATH OF HERSCHEL—SUMMER AT SORRENTO—BILL FOR PROTECTION OF ANIMALS—NINETY-SECOND YEAR—LETTER FROM PROFESSOR SEDGWICK—GRAND ERUPTION OF VESUVIUS—LAST SUMMER AT SORRENTO, PLANTS FOUND THERE—CONCLUSION.

THE summer of 1870 was unusually cool; but the winter has been extremely gloomy, with torrents of rain, and occasionally such thick fogs, that I could see neither to read nor to write. We had no storms during the hot weather; but on the afternoon of the 21st December, there was one of the finest thunderstorms I ever saw; the lightning was intensely vivid, and took the strangest forms, darting in all directions through the air before it struck, and sometimes darting from the ground or the sea to the clouds. It ended in a deluge of rain, which lasted all night, and made us augur ill for the solar eclipse next day; and, sure enough, when I awoke next morning, the sky was darkened by clouds and rain. Fortunately, it cleared up just as the eclipse began; we were all prepared for observing it, and we followed its progress through the opening in the clouds till at last there was only a very slender crescent of the sun's disc left; its convexity was turned upwards, and its horns were nearly horizontal. It was then hidden by a dense mass of clouds; but after a time they opened, and I saw the edge of the moon leave the limb of the sun. The appearance of the landscape was very lurid, but by no means very dark. The common people and children had a very good view of the eclipse, reflected by the pools of water in the streets.

Many of the astronomers who had been in Sicily observing the eclipse came to see me as they passed through Naples. One of their principal objects was to ascertain the nature of the corona, or bright white rays which surround the dark lunar disc at the time of the greatest obscurity. The spectroscope showed that it was decidedly auroral, but as the aurora was seen on the dark disc of the moon it must have been due to the earth's atmosphere. Part of the corona was polarized, and consequently must have been material; the question is, Can it be the etherial medium? A question of immense importance, since the whole theory of light and colours and the resistance of Encke's comet depends upon that hypothesis. The question is still in abeyance, but I have no doubt that it will be decided in the affirmative, and that even the cause of gravitation will be known eventually.

At this time I had the pleasure of a visit from Mr. Peirce, Professor of Mathematics and Astronomy, in the Harvard University, U.S., and Superintendent of the U.S. Coast Survey, who had come to Europe to observe the eclipse. On returning to America he kindly sent me a beautiful lithographed copy of a very profound memoir in linear and associative algebra. Although in writing my popular books I had somewhat neglected the higher algebra, I have read a great part of the work; but as I met with some difficulties I wrote to Mr. Spottiswoode, asking his advice as to the books that would be of use, and he sent me Serret's "Cours d'Algèbre Supérieure," Salmon's "Higher Algebra," and Tait on "Quaternions;" so now I got exactly what I wanted, and I am very busy for a few hours every morning; delighted to have an occupation so entirely to my mind. I thank God that my intellect is still unimpaired. I am grateful to Professor Peirce for giving me an opportunity of exercising it so agreeably. During the rest of the day I have recourse to Shakespeare, Dante, and more modern light reading, besides the newspapers, which always interested me much. I have resumed my habit of working, and can count the threads of a fine canvas without spectacles. I receive every one who comes to see me, and often have the pleasure of a visit from old friends very unexpectedly. In the evening I read a novel, but my tragic days are over; I prefer a cheerful conversational novel to the sentimental ones. I have recently been reading Walter Scott's novels again, and enjoyed the broad Scotch in them. I play a few games at Bézique with one of my daughters, for honour and glory, and so our evenings pass pleasantly enough.

It is our habit to be separately occupied during the morning, and spend the rest of the day together. We are fond of birds and have several, all very tame. Our tame nightingales sing very beautifully, but, strange to say, not at night. We have also some solitary sparrows, which are, in fact, a variety of the thrush (Turdus cyaneus), and some birds which we rescued from destruction in spring, when caught and ill-used by the boys in the streets; besides, we have our dogs; all of which afford me amusement and interest.

* * * * *

Mr. Murray has kindly sent me a copy of Darwin's recent work on the "Descent of Man." Mr. Darwin maintains his theory with great talent and with profound research. His knowledge of the characters and habits of animals of all kinds is very great, and his kindly feelings charming. It is chiefly by the feathered race that he has established his law of selection relative to sex. The males of many birds are among the most beautiful objects in nature; but that the beauty of nature is altogether irrelative to man's admiration or appreciation, is strikingly proved by the admirable sculpture on Diatoms and Foraminifera; beings whose very existence was unknown prior to the invention of the microscope. The Duke of Argyll has illustrated this in the "Reign of Law," by the variety, graceful forms and beautiful colouring of the humming birds in forests which man has never entered.

In Mr. Darwin's book it is amusing to see how conscious the male birds are of their beauty; they have reason to be so, but we scorn the vanity of the savage who decks himself in their spoils. Many women without remorse allow the life of a pretty bird to be extinguished in order that they may deck themselves with its corpse. In fact, humming birds and other foreign birds have become an article of commerce. Our kingfishers and many of our other birds are on the eve of extinction on account of a cruel fashion.

I have just received from Frances Power Cobbe

an essay, in which she controverts Darwin's theory,* so far as the origin of the moral sense is concerned. It is written with all the energy of her vigorous intellect as a moral philosopher, yet with a kindly tribute to Mr. Darwin's genius. I repeat no one admires Frances Cobbe more than I do. I have ever found her a brilliant, charming companion, and a warm, affectionate friend. She is one of the few with whom I keep up a correspondence.

To Mr. Murray I am indebted for a copy of Tylor's "Researches on the Early History of Mankind, and the Development of Civilization"—a very remarkable work for extent of research, original views, and happy illustrations. The gradual progress of the pre-historic races of mankind has laid a foundation from which Mr. Tylor proves that after the lapse of ages the barbarous races now existing are decidedly in a state of progress towards civilization. Yet one cannot conceive human beings in a more degraded state than some of them are still; their women are treated worse than their dogs. Sad to say, no savages are more gross than the lowest ranks in England, or treat their wives with more cruelty.

* * * * *

In the course of my life Paris has been twice occupied by foreign troops, and still oftener has

* "Darwinism in Morals," &c.

it been in a state of anarchy. I regret to see that La Place's house at Arcœuil has been broken into, and his manuscripts thrown into the river, from which some one has fortunately rescued that of the "Mécanique Céleste," which is in his own handwriting. It is greatly to the honour of French men of science that during the siege they met as usual in the hall of the Institute, and read their papers as in the time of peace. The celebrated astronomer Janssen even escaped in a balloon, that he might arrive in time to observe the eclipse of the 22nd November, 1870.

* * * * *

We had a most brilliant display of the Aurora on the evening of Sunday, the 4th February, 1871, which lasted several hours. The whole sky from east to west was of the most brilliant flickering white light, from which streamers of red darted up to the zenith. There was also a lunar rainbow. The common people were greatly alarmed, for there had been a prediction that the world was coming to an end, and they thought the bright part of the Aurora was a piece of the moon that had already tumbled down! This Aurora was seen in Turkey and in Egypt.

* * * * *

I am deeply grieved and shaken by the death of

Sir John Herschel, who, though ten years younger than I am, has gone before me. In him I have lost a dear and affectionate friend, whose advice was invaluable, and his society a charm. None but those who have lived in his home can imagine the brightness and happiness of his domestic life. He never presumed upon that superiority of intellect or the great discoveries which made him one of the most illustrious men of the age ; but conversed cheerfully and even playfully on any subject, though ever ready to give information on any of the various branches of science to which he so largely contributed, and which to him were a source of constant happiness. Few of my early friends now remain—I am nearly left alone.

* * * * *

We went to pass the summer and autumn at Sorrento, where we led a very quiet but happy life. The villa we lived in was at a short distance from and above the town, quite buried in groves of oranges and lemons, beyond which lay the sea, generally calm and blue, sometimes stormy ; to our left the islands of Ischia and Procida, the Capo Miseno, with Baia, Pozzuoli, and Posilipo ; exactly opposite to us, Naples, then Vesuvius, and all the little towns on that coast, and lastly, to our right, this wonderful panorama was bounded by the fine cliffs of the Monte Santangelo. It was beautiful always, but most beautiful when the sun, setting behind Ischia, sent a perfect glory over the rippling sea, and tinged the Monte Santangelo and the cliffs which bound the Piano di Sorrento literally with purple and gold. I spent the whole day on a charming terrace sheltered from the sun, and there we dined and passed the evening watching the lights of Naples reflected in the water and the revolving lights of the different lighthouses. I often drove to Massa till after sunset, for from that road I could see the island of Capri, and I scarcely know a more lovely drive. Besides the books we took with us we had newspapers, reviews, and other periodicals, so that we were never dull. On one occasion my daughters and I made an expedition up the hills to the Deserto, from whence one can see the Gulf of Salerno and the fine mountains of Calabria. My daughters rode and I was carried in a *portantina*. It was fine, clear, autumnal weather, and I enjoyed my expedition immensely, nor was I fatigued.

* * * * *

In November we returned to Naples, where I resumed my usual life. I had received a copy of Hamilton's Lectures on Quaternions from the Rev. Whitewell Elwin. I am not acquainted with that gentleman, and am the more grateful to him. I have now a valuable library of scientific books and transactions of scientific societies, the greater part gifts from the authors.

Foreigners were so much shocked at the atrocious cruelty to animals in Italy, that an attempt was made about eight years ago to induce the Italian Parliament to pass a law for their protection, but it failed. As Italy is the only civilized country in Europe in which animals are not protected by law, another attempt is now being made; I have willingly given my name, and I received a kind letter from the Marchioness of Ely, from Rome, to whom I had spoken upon the subject at Naples, telling me that the Princess Margaret, Crown Princess of Italy, had been induced to head the petition. Unless the educated classes take up the cause one cannot hope for much change for a long time. Our friend, Mr. Robert Hay, who resided at Rome for many years, had an old horse of which he was very fond, and on leaving Rome asked a Roman prince, who had very large possessions in the Campagna, if he would allow his old horse to end his days on his grassy meadows. "Certainly," replied the prince, "but how can you care what becomes of an animal when he is no longer of use?" We English cannot boast of humanity, however, as long as our sportsmen find pleasure in shooting down tame pigeons as they fly terrified out of a cage.

* * * * *

I am now in my 92nd year (1872), still able to drive out for several hours; I am extremely deaf, and my memory of ordinary events, and especially of the names of people, is failing, but not for mathematical and scientific subjects. I am still able to read books on the higher algebra for four or five hours in the morning, and even to solve the problems. Sometimes I find them difficult, but my old obstinacy remains, for if I do not succeed to-day, I attack them again on the morrow. I also enjoy reading about all the new discoveries and theories in the scientific world, and on all branches of science.

Sir Roderick Murchison has passed away, honoured by all, and of undying fame; and my amiable friend, almost my contemporary, Professor Sedgwick, has been obliged to resign his chair of geology at Cambridge, from age, which he had filled with honour during a long life.

[The following letter from her valued friend Professor Sedgwick, in 1869, is the last my mother received from him:—

FROM PROFESSOR SEDGWICK TO MRS. SOMERVILLE.

CAMBRIDGE, *April* 21*st*, 1869.

MY DEAR MRS. SOMERVILLE,

I heard, when I was in London, that you were still in good bodily health, and in full fruition of your great intellectual strength, while breathing the sweet air of Naples. I had been a close prisoner to my college rooms through the past winter and spring; but I broke from my prison-house at the beginning of this month, that I might consult my oculist, and meet my niece on her way to Italy. . . . . . . My niece has for many years (ever since 1840) been my loving companion during my annual turn of residence as canon of Norwich; and she is, and from her childhood has, been to me as a dear daughter. I know you will forgive me for my anxiety to hear from a living witness that you are well and happy in the closing days of your honoured life; and for my longing desire that my beloved daughter (for such I ever regard her) should speak to you face to face, and see (for however short an interview) the Mrs. Somerville, of whom I have so often talked with her in terms of honest admiration and deep regard. The time for the Italian tour is, alas! far too short. But it will be a great gain to each of the party to be allowed, even for a short time, to gaze upon the earthly paradise that is round about you, and to cast one look over its natural wonders and historic monuments. . . . . . . Since you were here, my dear and honoured guest, Cambridge is greatly changed. I am left here like a vessel on its beam ends, to mark the distance to which the current has been drifting during a good many bygone years. I have outlived nearly all my early friends. Whewell, Master of Trinity, was the last of the old stock who was living here. Herschel has not been here for several years. Babbage was here for a day or two during the year before last. The Astronomer-Royal belongs to a more recent generation. For many years long attacks of suppressed gout have made my life very unproductive. I yesterday dined in Hall. It was the first time I was able to meet my brother Fellows since last Christmas day. A long attack of bronchitis, followed by a distressing inflammation of my eyes, had made me a close prisoner for nearly four months. But, thank God, I am again beginning to be cheery, and with many infirmities (the inevitable results of old age, for I have entered on my 85th year) I am still strong in general health, and capable of enjoying, I think as much as ever, the society of those whom I love, be they young or old. May God preserve and bless you; and whensoever it may be His will to call you away to Himself, may your mind be without a cloud and your heart full of joyful Christian hope!

I remain,
My dear Friend,
Faithfully and gratefully yours,
ADAM SEDGWICK.

After all the violence and bloodshed of the preceding year, the Thanksgiving of Queen Victoria and the British nation for the recovery of the Prince of Wales will form a striking event in European history. For it was not the congregation in St.

Paul's alone, it was the spontaneous gratitude of all ranks and all faiths throughout the three kingdoms that were offered up to God that morning; the people sympathized with their Queen, and no sovereign more deserves sympathy.

* * * * *

Vesuvius has exhibited a considerable activity during the winter and early spring, and frequent streams of lava flowed from the crater, and especially from the small cone to the north, a little way below the principal crater. But these streams were small and intermittent, and no great outbreak was expected. On the 24th April a stream of lava induced us to drive in the evening to Santa Lucia. The next night, Thursday, 25th April, my daughter Martha, who had been to the theatre, wakened me that I might see Vesuvius in splendid eruption. This was at about 1 o'clock on Friday morning. Early in the morning I was disturbed by what I thought loud thunder, and when my maid came at 7 a.m. I remarked that there was a thunder storm, but she said, "No, no; it is the mountain roaring." It must have been very loud for me to hear, considering my deafness, and the distance Vesuvius is from Naples, yet it was nothing compared to the noise later in the day, and for many days after. My daughter, who had gone to Santa Lucia to see the eruption better, soon came to fetch me with our friend Mr. James Swinton, and we passed the whole day at windows in an hotel at Santa Lucia, immediately opposite the mountain. Vesuvius was now in the fiercest eruption, such as has not occurred in the memory of this generation, lava overflowing the principal crater and running in all directions. The fiery glow of lava is not very visible by daylight; smoke and steam is sent off which rises white as snow, or rather as frosted silver, and the mouth of the great crater was white with the lava pouring over it. New craters had burst out the preceding night, at the very time I was admiring the beauty of the eruption, little dreaming that, of many people who had gone up that night to the Atrio del Cavallo to see the lava (as my daughters had done repeatedly and especially during the great eruption of 1868), some forty or fifty had been on the very spot where the new crater burst out, and perished, scorched to death by the fiery vapours which eddied from the fearful chasm. Some were rescued who had been less near to the chasm, but of these none eventually recovered.

Behind the cone rose an immense column of dense black smoke to more than four times the height of the mountain, and spread out at the summit horizontally, like a pine tree, above the silvery stream which poured forth in volumes. There were constant bursts of fiery projectiles, shooting to an immense height into the black column of smoke, and tinging it with a lurid red colour. The fearful roaring and thundering never ceased for one moment, and the house shook with the concussion of the air. One stream of lava flowed towards Torre del Greco, but luckily stopped before it reached the cultivated fields; others, and the most dangerous ones, since some of them came from the new craters, poured down the Atrio del Cavallo, and dividing before reaching the Observatory flowed to the right and to the left—the stream which flowed to the north very soon reached the plain, and before night came on had partially destroyed the small town of Massa di Somma. One of the peculiarities of this eruption was the great fluidity of the lava; another was the never-ceasing thundering of the mountain. During that day we observed several violent explosions in the great stream of lava: we thought from the enormous volumes of black smoke emitted on these occasions that new craters had burst out—some below the level of the Observatory; but that can hardly have been the case. My daughters at night drove to Portici, and went up to the top of a house, where the noise seems to have been appalling; but they told me they did not gain anything by going to Portici, nor did they see the eruption better than I did who remained at Santa Lucia, for you get too much below the mountain on going near. On Sunday, 28th, I was surprised at the extreme darkness, and on looking out of window saw men walking with umbrellas; Vesuvius was emitting such an enormous quantity of ashes, or rather fine black sand, that neither land, sea, nor sky was visible; the fall was a little less dense during the day, but at night it was worse than ever. Strangers seemed to be more alarmed at this than at the eruption, and certainly the constant loud roaring of Vesuvius was appalling enough amidst the darkness and gloom of the falling ashes. The railroad was crowded with both natives and foreigners, escaping; on the other hand, crowds came from Rome to see the eruption. We were not at all afraid, for we considered that the danger was past when so great an eruption had acted as a kind of safety-valve to the pent-up vapours. But a silly report got about that an earthquake was to take place, and many persons passed the night in driving or walking about the town, avoiding narrow streets. The mountain was quite veiled for some days by vapour and ashes, but I could see the black smoke and silvery mass above it. While looking at this, a magnificent column, black as jet, darted with inconceivable violence and velocity to an immense height; it gave a grand idea of the power that was still in action in the fiery caverns below.

Immense injury has been done by this eruption, and much more would have been done had not the lava flowed to a great extent over that of 1868. Still the streams ran through Massa di Somma, San Sebastiano, and other villages scattered about the country, overwhelming fields, woods, vineyards, and houses. The ashes, too, have not only destroyed this year's crops, but killed both vines and fruit trees, so that altogether it has been most disastrous. Vesuvius was involved in vapour and ashes till far on in May, and one afternoon at sunset, when all below was in shade, and only a few silvery threads of steam were visible, a column of the most beautiful crimson colour rose from the crater, and floated in the air. Many of the small craters still smoked, one quite at the base of the cone, which is a good deal changed—it is lower, the small northern cone has disappeared, and part of the walls of the crater have fallen in, and there is a fissure in them through which smoke or vapour is occasionally emitted.

* * * * *

On the 1st June we returned to Sorrento, this time to a pretty and cheerful apartment close to the sea, where I led very much the same pleasant life as the year before—busy in the morning with my own studies, and passing the rest of the day on the terrace with my daughters, who brought me beautiful wild flowers from their excursions over the country. Many of the flowers they brought were new to me, and it is a curious fact that some plants which did not grow in this part of the country a few years ago are now quite common. Amongst others, the Trachelium cœruleum, a pretty wall-plant, native of Calabria, and formerly unknown here, now clothes many an old wall near Naples, and at Sorrento. The ferns are extremely beautiful here. Besides those common to England, the Pteris cretica grows luxuriantly in the damp ravines, as well as that most beautiful of European ferns, the Woodwardia radicans, whose fronds are often more than six feet long. The inhabitants of Sorrento are very superior to the Neapolitans, both in looks and character; they are cleanly, honest, less cruel to animals, and have pleasant manners—neither too familiar nor cringing. As the road between Sorrento and Castellamare was impassable, owing to the fall of immense masses of rock from the cliffs above it, we crossed over in the steamer with our servants and our pet birds, for I now have a beautiful long-tailed parroquet called Smeraldo, who is my constant companion and is very familiar

And here I must mention how much I was pleased to hear that Mr. Herbert, M.P., has brought in a bill to protect land birds, which has been passed in Parliament; but I am grieved to find that "The lark which at Heaven's gate sings" is thought unworthy of man's protection. Among the numerous plans for the education of the young, let us hope that mercy may be taught as a part of religion.

*   *   *   *   *

Though far advanced in years, I take as lively an interest as ever in passing events. I regret that I shall not live to know the result of the expedition to determine the currents of the ocean, the distance of the earth from the sun determined by the transits of Venus, and the source of the most renowned of rivers, the discovery of which will immortalise the name of Dr. Livingstone. But I regret most of all that I shall not see the suppression of the most atrocious system of slavery that ever disgraced humanity—that made known to the world by Dr. Livingstone and by Mr. Stanley, and which Sir Bartle Frere has gone to suppress by order of the British Government.

*   *   *   *   *

The Blue Peter has been long flying at my foremast, and now that I am in my ninety-second year I must soon expect the signal for sailing. It is a solemn voyage, but it does not disturb my tranquillity. Deeply sensible of my utter unworthiness, and profoundly grateful for the innumerable blessings I have received, I trust in the infinite mercy of my Almighty Creator. I have every reason to be thankful that my intellect is still unimpaired, and, although my strength is weakness, my daughters support my tottering steps, and, by incessant care and help, make the infirmities of age so light to me that I am perfectly happy.

I HAVE very little more to add to these last words of my Mother's Recollections. The preceding pages will have given the reader some idea—albeit perhaps a very imperfect one—of her character and opinions. Only regarding her feelings on the most sacred of themes, is it needful for me to say a few words. My mother was profoundly and sincerely religious; hers was not a religion of mere forms and doctrines, but a solemn deep-rooted faith which influenced every thought, and regulated every action of her life. Great love and reverence towards God was the foundation of this pure faith, which accompanied her from youth to extreme old age, indeed to her last moments, which gave her strength to endure many sorrows, and was the mainspring of that extreme humility which was so remarkable a feature of her character.

At a very early age she dared to think for herself, fearlessly shaking off those doctrines of her early creed which seemed to her incompatible with the unutterable goodness and greatness of God; and through life she adhered to her simple faith, holding quietly and resolutely to the ultimate truths of religion, regardless alike of the censure of bigots or the smiles of sceptics. The theories of modern science she welcomed as quite in accordance with her religious opinions. She rejected the notion of occasional interference by the Creator with His work, and believed that from the first and invariably He has acted according to a system of harmonious laws, some of which we are beginning faintly to recognise, others of which will be discovered in course of time, while many must remain a mystery to man while he inhabits this world. It was in her early life that the controversy raged respecting the incompatibility of the Mosaic account of Creation, the Deluge, &c., with the revelations of geology. My mother very soon accepted the modern theories, seeing in them nothing in any way hostile to true religious belief. It is singular to recall that her candid avowal of views now so common, caused her to be publicly censured by name from the pulpit of York Cathedral. She foresaw the great modifications in opinion which further discoveries will inevitably produce; but she foresaw them without doubt or fear. Her constant prayer was for light and truth, and its full accomplishment she looked for confidently in the life beyond the grave. My mother never discussed religious subjects in general society; she considered them far too solemn to be talked of lightly; but with those near and dear to her, and with very intimate friends, whose opinion agreed with her own, she spoke freely and willingly. Her mind was constantly occupied with thoughts on religion; and in her last years especially she reflected much on that future world which she expected soon to enter, and lifted her heart still more frequently to that good Father whom she had loved so fervently all her life, and in whose merciful care she fearlessly trusted in her last hour.

My mother's old age was a thoroughly happy one. She often said that not even in the joyous spring of life had she been more truly happy. Serene and cheerful, full of life and activity, as far as her physical strength permitted, she had none of the infirmities of age, except difficulty in hearing, which prevented her from joining in general conversation. She had always been near-sighted, but could read small print with the greatest ease without glasses, even by lamp-light. To the last her intellect remained perfectly unclouded; her affection for those she loved, and her sympathy for all living beings, as fervent as ever; nor did her ardent desire for and belief in the ultimate religious and moral improvement of mankind diminish. She always retained her habit of study, and that pursuit, in which she had attained such excellence and which was always the most congenial to her, —Mathematics—delighted and amused her to the end. Her last occupations, continued to the actual day of her death, were the revision and completion of a treatise, which she had written years before, on the "Theory of Differences" (with diagrams exquisitely drawn), and the study of a book on Quaternions. Though too religious to fear death, she dreaded outliving her intellectual powers, and it was with intense delight that she pursued her intricate calculations after her ninetieth and ninety-first years, and repeatedly told me how she rejoiced to find that she had the same readiness and facility in comprehending and developing these extremely difficult formulæ which she possessed when young. Often, also, she said how grateful she was to the Almighty Father who had allowed her to retain her faculties unimpaired to so great an age. God was indeed loving and merciful to her; not only did He spare her this calamity, but also the weary trial of long-continued illness. In health of body and vigour of mind, having lived far beyond the usual span of human life, He called her to Himself. For her Death lost all its terrors. Her pure spirit passed away so gently that those around her scarcely perceived when she left them. It was the beautiful and painless close of a noble and a happy life.

My mother died in sleep on the morning of the 29th Nov., 1872. Her remains rest in the English Campo Santo of Naples.

THE END.

Printed in the United States
By Bookmasters